# Renal Ammoniagenesis and Interorgan Cooperation in Acid-Base Homeostasis

# Contributions to Nephrology

**Vol. 110**



KARGER

Basel · Freiburg · Paris · London · New York ·
New Delhi · Bangkok · Singapore · Tokyo · Sydney

6th International Workshop on Ammoniagenesis,
Mortola/Ventimiglia, June 8–11, 1993

# Renal Ammoniagenesis and Interorgan Cooperation in Acid-Base Homeostasis

Volume Editors

*A. Tizianello,* Genoa
*G. Baverel,* Lyon
*H. Endou,* Tokyo
*A.C. Schoolwerth,* Richmond, Va.
*D.J. O'Donovan,* Galway

63 figures and 25 tables, 1994

KARGER

Basel · Freiburg · Paris · London · New York ·
New Delhi · Bangkok · Singapore · Tokyo · Sydney

.............................

# Contributions to Nephrology

Library of Congress Cataloging-in-Publication Data
International Workshop on Ammoniagenesis (6th: 1993: Mortola, Italy)
Renal ammoniagenesis and interorgan cooperation in acid-base homeostasis: 6th International Workshop on Ammoniagenesis, Mortola/Ventimiglia, June 8–11, 1993 / volume editors, A. Tizianello ... (et al.).
(Contributions to nephrology; v. 110)
Includes bibliographical references and index.
1. Kidneys – Metabolism – Congresses. 2. Ammonia – Metabolism – Congresses.
3. Glutamine – Metabolism – Congresses. 4. Acid-base equilibrium – Congresses.
I. Tizianello, A. (Alberto), 1925– . II. Title. III. Series.
QP249.I567 1993 612.4'63--dc20
ISBN 3-8055-5915-1

Bibliographic Indices. This publication is listed in bibliographic services, including Current Contents® and Index Medicus.

Drug Dosage. The authors and the publisher have exerted every effort to ensure that drug selection and dosage set forth in this text are in accord with current recommendations and practice at the time of publication. However, in view of ongoing research, changes in government regulations, and the constant flow of information relating to drug therapy and drug reactions, the reader is urged to check the package insert for each drug for any change in indications and dosage and for added warnings and precautions. This is particularly important when the recommended agent is a new and/or infrequently employed drug.


Printed in Switzerland on acid-free paper by Thür AG Offsetdruck, Pratteln
ISBN 3-8055-5915-1

# Contents



## Ammonia-Forming Enzymes and Regulation of Renal Ammoniagenesis

## Renal Ammonia Transport



## Ammonia Production and Substrate Utilization along the Nephron



## Extrarenal Metabolism and Transport of Ammonia and Substrates

## Interorgan Metabolism of Ammonia and Substrates in Acid-Base Homeostasis

# Dedication to Dr. Guy Lemieux

The First International Workshop on Ammoniagenesis was jointly organized by Guy Lemieux, Arnold Relman and Patrick Vinay and was held near Montreal in June 1978 as a satellite of the VIIth International Congress of Nephrology. Since this initial event, a similar ammoniagenesis workshop has taken place as a satellite of each congress of the International Society of Nephrology. Meetings on ammonia have been held in Greece, the USA, France and Japan, and were always an exceptional occasion to meet fellow scientists, to exchange results and ideas, and to enrich everyone's understanding by the expertise and interactions of all participants. It has not only been a scientific success, but also a story of friendship and sharing. The proceedings of the meetings have been published regularly ever since the second reunion. We are fortunate to meet again in Italy today, before the Jerusalem International Congress. It is appropriate to dedicate the present meeting to one of the founding fathers of this Society who, since the last reunion, is no longer among us. Guy Lemieux died in 1991. He is not only remembered as an enthusiastic colleague, passionate in life, but also as an outstanding researcher in the field of urinary acidification and ammonia. A man of great stature, he had many interactions with several of the scientists present here. He was a dynamic and meticulous researcher. He trained a large number of scientists and nephrologists. He will be remembered as one of the Canadian pioneers in the field of ammonia. Science grows like a crystal by the creation of multiple sparkles of knowledge. There is no doubt that the contribution of Guy Lemieux was one of those sparkles and not the least of them in this very specialized field.

I think that we can all say that we are proud of contributing to the growth of this crystal and that we have found a marvellous brotherhood with many people of great quality that we would not have met without sharing the same interest. Let us address our thanks especially to Guy Lemieux and to former colleagues who are no longer active in this field. Let us hope that new contributors will continue to join us in the unravelling of the wonderful world of metabolic and biochemical regulators of the kidney.

*Patrick Vinay*, MD

# Preface

This issue is comprised of papers presented at the 6th International Workshop on Ammoniagenesis held June 8–11 in Villa Hanbury, Mortola, Italy. The presentations were classified into five areas dealing with: (1) ammonia-forming enzymes and regulation of renal ammoniagenesis; (2) renal ammonia transport; (3) ammonia production and substrates utilization along the nephron; (4) extrarenal metabolism and transport of ammonia and substrates, and (5) interorgan metabolism of ammonia and substrates in acid-base homeostasis. However, the categorization is quite arbitrary, and overlapping exists among the publications in the various sections.

The Organizing Committee would like to express its gratitude to the Consiglio Nazionale delle Ricerche and in addition to Baxter, Bayer Italia, Bellco, Boehringer-Mannheim Italia, Cilag, Emac, Gambro, Gruppo Lepetit, Hoechst Italia, Janssen Farmaceutici, Merck Sharp & Dohme, Menarini, Pfizer Italiana, REN Corporation, USA, Sandoz Prodotti Farmaceutici, Società Italiana Farmaceutici Ravizza for their financial support to the organization of this Workshop and publication of the proceedings.

Finally, the Organizing Committee gratefully acknowledges the University of Genoa which offered hospitality for the scientific sessions in the Hanbury Botanical Institute, and the Chief and Staff of this Institute for their wonderful cooperation.

*A. Tizianello*
*G. Baverel*
*H. Endou*
*A.C. Schoolwerth*
*D.J. O'Donovan*

Tizianello A, Baverel G, Endou H, Schoolwerth AC, O'Donovan DJ (eds):
Renal Ammoniagenesis and Interorgan Cooperation in Acid-Base Homeostasis.
Contrib Nephrol. Basel, Karger, 1994, vol 110, pp 1–6

# Glucocorticoid Receptor Homeostasis and Renal Glutaminase Activity

*M. Nam, T.C. Welbourne*[1]

Department of Physiology, Louisiana State University Medical Center,
Shreveport, La., USA

Chronic glucocorticoid (Gc) administration accelerates both the rate of glutamine synthesis and hydrolysis [1, 2]. Although the former is clearly associated with increased amounts of glutamine synthetase [1, 2], the increased glutamine breakdown is puzzling given the glutaminase response: acutely high doses depress renal glutaminase [3] whereas chronically the same dose has the opposite effect [1, 2]. Since Gc act to down-regulate glucocorticoid receptor (GR) expression [4], these differences may reflect the relationship between the functional GR population and the glutaminase gene expression. The following studies were designed to test this hypothesis.

## Materials and Methods

Our strategy was to manipulate the GR population in vivo through ligand availability and hence autoregulation of the GR gene and in vitro by bolstering the endogenous population by introducing the GR gene into proximal tubule-like cultured kidney cells.

In vivo experiments were performed on male Sprague-Dawley rats weighing 300–350 g. Adrenalectomies (ADX) were performed via the dorsal approach (Harlan, Indianapolis, Ind., USA) with the animals maintained on 0.9% NaCl prior to study. Experiments were always carried out within 12 days of ADX. Gc was administered in the form of a subdermal pellet releasing triamcinolone at a constant daily rate (Innovative Products, Toledo, Ohio, USA) of either 0.40 or 40 μg/100 g b.w., low or high dose, respectively.

[1] The excellent secretarial support of Tracey L. Dawson is gratefully acknowledged. Special thanks to Drs. John Cidlowski and Norm Curthoys for their support. Studies supported by DK-21026 and by the Louisiana Quality Education Support Fund RD-A-15.

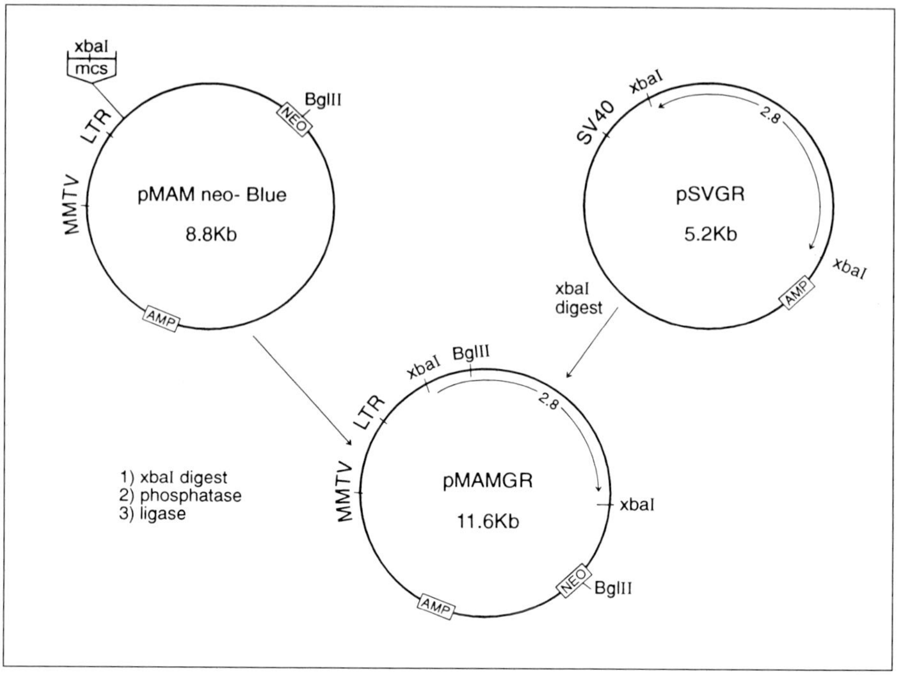


*Fig. 1.* GR expression vector, pMAMGR. pMAMGR contains the 2.8-kb rat GR gene XbaI cut of pSVGR (kindly provided by Dr. Roger Miesfeld) fused downstream of the MMTV promoter driving the pMAM neo-Blue expression vector (Clontech, Palo Alto, Calif., USA). The XbaI cut fragment was ligated into the multiple cloning site (MCS) after XbaI digestion and phosphatase treatment. The plasmid was grown up in competent HB10 cells in the presence of ampicillin and harvested off Qiagen columns (Qiagen, Chatsworth, Calif., USA).

After 24 h (acute) or 72 h (chronic) administration, the animals were anesthetized and their kidneys removed for cytosolic receptor analysis [1] and glutaminase and synthetase activity [1].

In vitro experiments were performed on LLCPK-$F^+$ cells [5] grown on modified DMEM media containing 5 m*M* glucose and 2 m*M* *L*-glutamine. At approximately 70% confluency the cells were transfected (Lipofectin; Life Technologies, Gaithersburg, Md., USA) with pMAMGR (see figure 1 for construct) and pGMCS which contains the chloramphenicol acetyltransferase (CAT) gene fused downstream of the mouse mammary tumor virus promotor and glucocorticoid regulatory element. After obtaining confluency, usually 48 h, the transfected cells were exposed to either vehicle alone or vehicle plus $10^{-6}$ *M* dexamethasone for an additional 24 h. The cells were then harvested and assayed for CAT activity [6] and glutaminase activity [1] with protein determined by a dye-binding assay (Sigma, St. Louis, Mo., USA) using albumin as the standard.

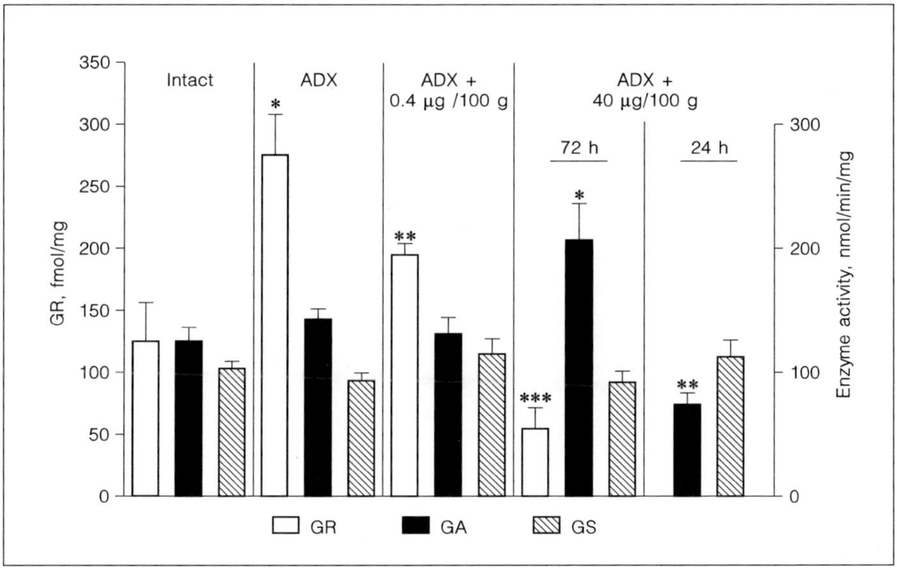


*Fig. 2.* Kidney cytosolic GR population and glutaminase and glutamine synthetase activity as a function of glucocorticoid availability. Adrenalectomized rats were sham treated or administered triamcinolone at low (0.4 μg/100 g b.w.) or high (40 μg/100 g b.w.) dose for 72 h or for 24 h (only high dose). Results are means ± SEM from 4 to 6 rats for group with statistical analysis by analysis of variance and posttest analysis employing a corrected t test (Student-Newman-Keuls). * Different from intact, $p < 0.05$. ** Different from ADX, $p < 0.05$. *** Different from ADX + 0.4 μg/100 g, $p < 0.05$.

## Results

*In vivo.* Rat kidney cytosolic receptor content (fig. 2) was manipulated over a 5-fold range, 54–275 fmol/mg protein by ADX and ADX plus high triamcinolone dose, 40 μg/100 g b.w. Note that the rise in cytosolic receptors with ADX may reflect receptor up-regulation [7] or nuclear GR migration; in contrast, the reduced number at high Gc reflects down-regulation of the receptor gene expression [4]. Acutely (24 h) high Gc (40 μg/100 g b.w.) reduced glutaminase activity (42%) compared to intact animals, consistent with our previous findings [3], and suggested that glutaminase like the GR may be negatively regulated. Chronically (72 h), the high dose Gc was associated with a markedly reduced receptor population (57%), compared to intact animals, and an elevated glutaminase activity (64%), suggesting that glutamine hydrolysis may be inversely related to the number of functional receptors.

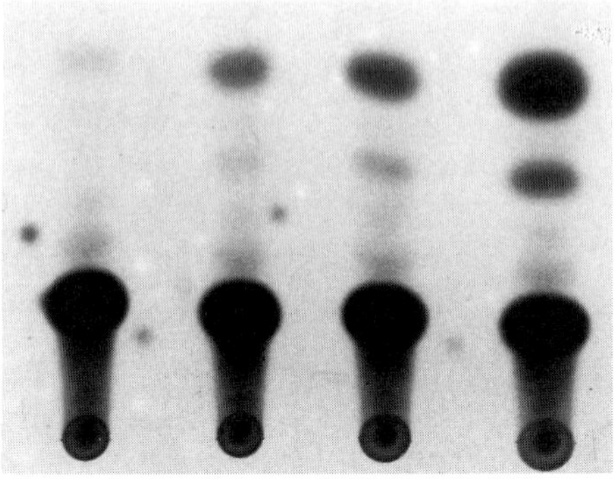


*Fig. 3.* Expression of CAT activity in LLCPK-F+ cells transfected with pGMCS and cotransfected with pMAMGR after exposure to dexamethasone (DEX) $10^{-6}$ *M*. pMAMGR was transfected using 1 and 2 μg of DNA (0.5 μg/ml). CAT activity was quantitated as percent conversion of chloramphenicol to its acetylated forms.

*In vitro.* LLCPK-F+ cells expressed a variable but generally low number of cytosolic receptors ranging from 56 to 102 fmol/mg protein. Since we wished to establish effects of high receptor complexes [8] on glutaminase activity, functional receptors were increased by introducing the GR expressing vector pMAMGR; that this did in fact occur is supported by results shown in figure 3. Cells transfected with the glucocorticoid-responsive CAT gene alone exhibited a 5-fold increase in CAT activity, 0.11–0.55% in the presence of dexamethasone, indicating the presence of functional receptors in these cells. Cotransfecting with pMAMGR increased the CAT activity 12.5- and 44-fold and therefore demonstrated the expression of additional functional receptors. The CAT assay results from 5 experiments and the corresponding glutaminase activity are presented in figure 4. In the absence of dexamethasone, glutaminase activity was $23 \pm 5$ nmol/min/mg protein and CAT activity only $0.08 \pm 0.01$%. Exposure to dexamethasone results in a small rise in CAT activity to $0.31 \pm 0.08$% while glutaminase was unchanged; this reflected a rise in activity in 2 experiments and a fall in the remainder. Interestingly, this rise in glutaminase activity occurred in experiments with the smallest increase in CAT activity. Cells transfected with pMAMGR expressed a large increase in CAT activity

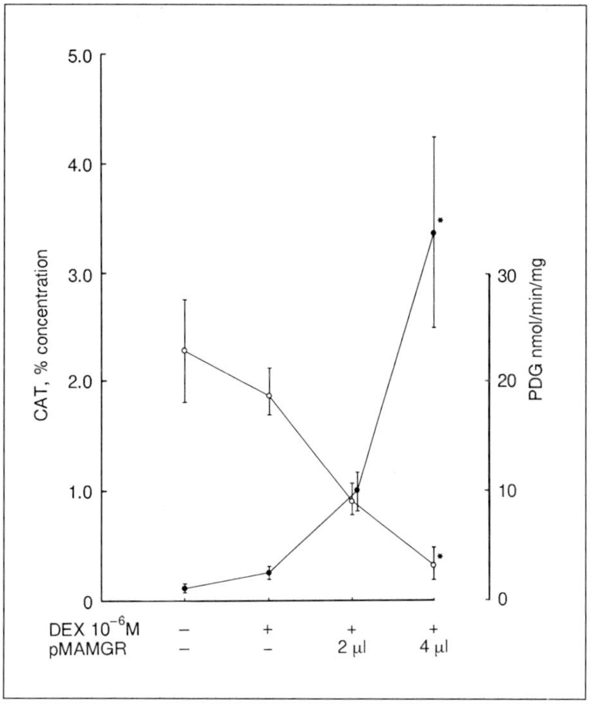


*Fig. 4.* Relationship between CAT activity expression and glutaminase activity in LLCPK-F+ cells cotransfected with pGMCS and pMAMGR. Results are means ± SEM from 5 experiments and analyzed by analysis of variance with significant differences, * $p < 0.05$ identified using a corrected t test (Bonferroni).

compared to the nontransfected cells, 3.3- and 11-fold, associated with a marked fall in glutaminase activity, 62 and 73% respectively. The failure to completely suppress glutaminase presumably reflects the limits of transfection efficiency.

## Discussion

Our goal in elucidating hormonal effects on interorgan glutamine flow is to define events at the molecular level and to resolve them in terms of meaningful physiological conditions. At first glance the role of glucocorticoid on organ as well as whole animal glutamine fluxes is paradoxical: acutely the animals are in an anabolic state whereas chronically they shift to a decided catabolic state. Associating these two conditions with glutamine fluxes suggests that net glutamine synthesis of the former reflects elevated glutamine synthetase and suppressed glutaminase expression while net glutamine breakdown of the latter reflects rebounding glutaminase activity. If so, the GR population should be seen as playing a major regulatory role in directing interorgan glutamine flow.

A two-tiered mechanism dependent upon GR number for controlling gene expression would not be unique for the glutaminase gene. Other genes, for example growth hormone, are regulated by Gc-activated receptor complexes at both the transcriptional and posttranscriptional levels [9]. What makes the glutaminase response so intriguing is the possibility of high level receptor complexes acting negatively through one limb and the low level complexes acting positively on the other limb.

## References


1 Tamarappoo B, Nam M, Kilberg M, Welbourne T: Glucocorticoid regulation of splanchnic glutamine, alanine, glutamate, ammonia, and glutathione fluxes. Am J Physiol 1993;264: E526–E533.

2 Welbourne TC: Glucocorticoid regulation of renal glutamate metabolism. Contrib Nephrol. Basel, Karger, 1991, vol 92, pp 52–56.

3 Welbourne C, Welbourne T: Effect of glucocorticoid on ammoniagenesis and glutamine utilizing enzymes of adrenalectomized rat kidney. Physiologist 1982;25:295A.

4 Burnstein K, Cidlowski J: The down side of glucocorticoid receptor regulation. Mol Cell Endocrinol 1992;83:51–58.

5 Gstraunthaler G, Handler J: Isolation, growth, and characterization of a gluconeogenic strain of renal cells. Am J Physiol 1987;252:C232–C238.

6 Gorman C, Moffat L, Howard B: Recombinant genomes which express chloramphenicol acetyltransferase in mammalian cells. Mol Cell Biol 1982;2:1044–1051.

7 Claire M, Oblin M-E, Steimer J-L, Nakame H, Misumi J, Michaud D, Corvol P: Effect of adrenalectomy and aldosterone on the modulation of mineralocorticoid receptors in rat kidney. J Biol Chem 1981;256:142–147.

8 Vanderbilt J, Miesfeld R, Maier B, Yamamoto K: Intracellular receptor concentration limits glucocorticoid-dependent enhancer activity. Mol Endocrinol 1987;1:68–94.

9 Diamond D, Goodman H: Regulation of growth hormone messenger RNA synthesis by dexamethasone and triiodothyronine: Transcriptional rate and mRNA stability changes in pituitary tumor cells. Mol Biol 1985;181:41–62.



T. C. Welbourne, PhD, Department of Physiology,
PO Box 33932, LSUMC, Shreveport, LA 71130 (USA)

Tizianello A, Baverel G, Endou H, Schoolwerth AC, O'Donovan DJ (eds):
Renal Ammoniagenesis and Interorgan Cooperation in Acid-Base Homeostasis.
Contrib Nephrol. Basel, Karger, 1994, vol 110, pp 7–12

# Properties of the Pig Renal Mitochondrial Glutamine Transporter

*B. Roberg, I. Aa. Torgner, E. Kvamme*

Neurochemical Laboratory, Preclinical Medicine, University of Oslo, Blindern, Oslo, Norway

Glutamine is important for renal ammonia production and to some extent for the energy metabolism [1]. Previous studies provide evidence for a specific renal mitochondrial glutamine transporter in the pig and rat, where a neutral uniport process has been suggested [2–5].

In this report we have further characterized the pig renal mitochondrial glutamine transport and have shown that it is saturable, largely stereospecific for the *L*-form, and stimulated by lowering the matrix pH. The uptake is complex and may occur by different channels. Thus, it is stimulated by $K^+$ and by raising the electrochemical potential across the inner mitochondrial membrane, with the implication of both the $\Delta$pH and the membrane potential ($\Delta\psi$).

## Methods

The mitochondria were isolated from fresh pig kidney cortex by differential centrifugation [6]. Glutamine transport was monitored as previously described, either by glutamine-induced swelling of mitochondria [6, 7] or by the uptake of [$^3$H]glutamine into the mitochondria [4, 5] at 25 °C. The two methods complement each other. The swelling was performed in an iso-osmotic solution containing 0.20 *M* glutamine and the decrease in absorbance at 546 nm measured, and corrected for the decrease with no glutamine present. Although semiquantitative, this method minimizes interference by glutaminase because of the large concentration of glutamine used, but some of the uptake is probably due to diffusion.

The uptake into the mitochondria of $10^{-4}$ *M* [$^3$H]glutamine (when not indicated otherwise) was monitored at 15 s, starting with the addition of glutamine, and the mitochondria were separated by rapid filtration. The radioactivity measured on the filter was corrected for binding to the filter itself. Short incubation times and low substrate concentration were used to minimize diffusion and interference by glutaminase. The

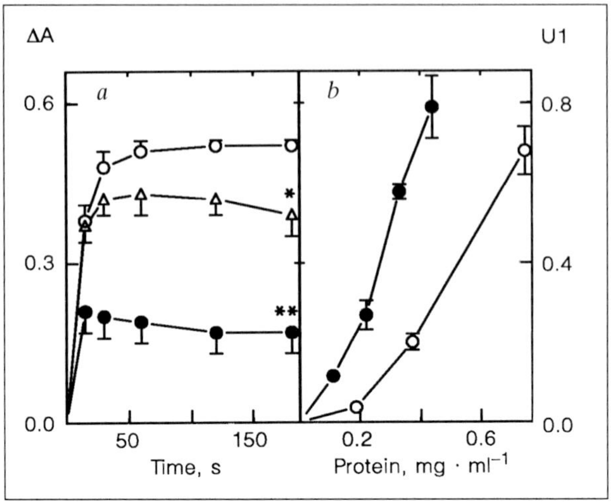


*Fig. 1.* Glutamine-induced swelling and [$^3$H]glutamine uptake in the mitochondria. *a* Time course of glutamine-induced swelling in *L*-glutamine, control (○); *L*-glutamine in the presence of 0.032 *M* *L*-glutamate (△), and *D*-glutamine substituted for *L*-glutamine (●). Data are mean ± SEM, n = 3. Where no error bars are shown, they are contained within the limits of the symbols. *b* The effect of protein concentration on glutamine-induced swelling, n = 3 (●) ΔA (left axis), and [$^3$H]glutamine uptake, n = 4 (○). U1: nmol · min$^{-1}$· ml assay volume$^{-1}$.

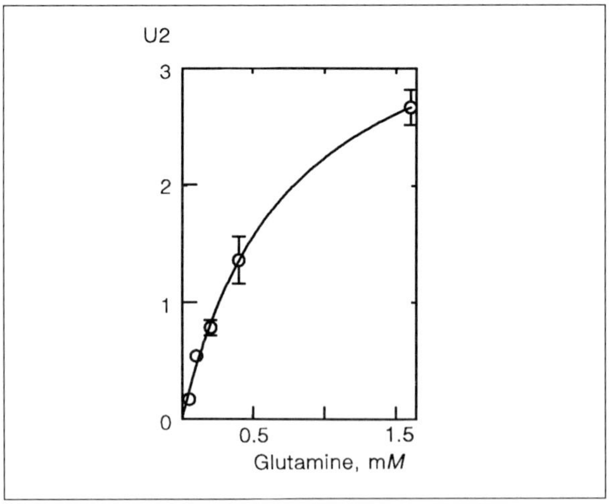


*Fig. 2.* The effect of the glutamine concentration on [$^3$H]glutamine uptake, n = 4. U2: nmol · min$^{-1}$· mg protein$^{-1}$.

*Table 1.* Effect of pH and ionophores on glutamine-induced swelling

| Experiment | KCl m*M* | $\Delta A_{546}$ (n) ± SE | | | |
|---|---|---|---|---|---|
| | | pH 6.5 | | pH 8.0 | |
| | | control | ionophore | control | ionophore |
| I | 32 | 0.47 (3) ± 0.02 | 0.37 (4) ± $0.01^{a}$ | 0.25 (3) ± $0.02^{b}$ | 0.16 (3) ± $0.03^{c}$ |
| II | 32 | 0.51 (6) ± 0.03 | 0.36 (6) ± $0.03^{a}$ | 0.17 (3) ± $0.02^{b}$ | 0.21 (3) ± 0.04 |
| III | 0 | 0.40 (5) ± 0.01 | 0.31 (5) ± $0.01^{a}$ | 0.21 (3) ± $0.03^{b}$ | 0.19 (3) ± $0.01^{c}$ |

Ionophores: I, FCCP, $10^{-4}$ *M*; II and III, nigericin, $2 \cdot 10^{-6}$ *M*.
$^{a}$ $p < 0.01$ vs. control, pH 6.5; $^{b}$ $p < 0.001$ vs. control, pH 6.5; $^{c}$ $p < 0.01$ vs. ionophore, pH 6.5.

measured uptake represents the sum of glutamine and its metabolites, but under our conditions we observed only about 1% conversion of the substrate. The nonspecific binding to mitochondria could amount to about 20–30%. The protein concentration ranged within 0.2–0.7 $mg \cdot ml^{-1}$ as determined by the method of Lowry et al. [8]. The data are expressed as mean ± statistical error of means (SE). Statistical significance was tested by Student's unpaired, two-tailed t test (* $p < 0.05$; ** $p < 0.01$).

## Results

The glutamine-induced swelling in *D*-glutamine is about 30% of that in *L*-glutamine, and *L*-glutamate inhibits the swelling slightly (fig. 1a). The other product of the glutaminase reaction, ammonia, has no effect (not shown). The uptake is positively, but nonlinearly correlated with the protein concentration, measured both with swelling and the filter method (fig. 1b). Furthermore, it is saturable, as demonstrated by a plot of uptake versus [$^{3}$H]glutamine concentration (fig. 2).

The ionophores carbonyl cyanide *p*-trifluoromethoxyphenylhydrazide (FCCP), which acts to dissipate the electrochemical proton potential, and nigericin, a $K^{+}/H^{+}$ exchanger, exert a pronounced inhibitory effect on glutamine-induced swelling at pH 6.5, but not at pH 8.0 (table 1). The swelling is also markedly stimulated by lowering the pH from 8.0 to 6.5 (table 1), in the presence of FCCP and nigericin as well. Furthermore, valinomycin that reduces the

*Table 2.* Effect of KCl and ionophores on glutamine-induced swelling

| Experiment | $\Delta A_{546}$ (n) ± SE | | | | | |
|---|---|---|---|---|---|---|
| | KCl | NaCl | Tris/HCl | Tris/Hepes | KCl+Val | KCl+Nig |
| I | 0.69 (6)<br>±0.02 | 0.37 (6)<br>±0.02[a] | 0.31 (3)<br>±0.03[a, d] | 0.45 (3)<br>±0.02[a] | 0.42 (4)<br>±0.03[a] | 0.40 (4)<br>±0.03[a] |
| | KCl | KCl+<br>Rot | KCl+Rot<br>+Val | KCl+Rot<br>+Nig | KCl+Olig<br>+Ant | KCl+Olig<br>+Ant+Val |
| II | 0.59 (3)<br>±0.02 | 0.35 (3)<br>±0.02[a] | 0.29 (3)<br>±0.01[b] | 0.25 (3)<br>±0.02 | 0.33 (3)<br>±0.01[a] | 0.27 (3)<br>±0.01[c] |

[a] $p < 0.001$ vs. KCl; [b] $p < 0.05$ vs. KCl + rotenone; [c] $p < 0.05$ vs. KCl+oligomycin+antimycin A; [d] $p < 0.05$ vs. Tris/Hepes.

pH 7.4; KCl, NaCl, Tris/HCl, Tris/Hepes: 0.032 *M*; valinomycin (Val): $10^{-5}$ *M*; nigericin (Nig): $2 \cdot 10^{-7}$ *M*; oligomycin (Olig): 10 mg · $l^{-1}$; antimycin A (Ant): 0.06 mg · $l^{-1}$; rotenone (Rot): $1.5 \cdot 10^{-5}$ *M*.

membrane potential in the presence of $K^+$, inhibits the swelling (table 2). It is stimulated by $K^+$, but $Na^+$ has no effect when compared to Tris/HCl or Tris/-Hepes, whereas $Cl^-$ has a minor, but significant inhibitory effect. When the metabolic inhibitors rotenone or oligomycin plus antimycin A are present, the swelling is strongly inhibited. However, valinomycin exerts an additional inhibition.

## Discussion

We have previously reported that glutamine transport into intact pig renal mitochondria is inhibited both by mersalyl, $Ca^{2+}$ and alanine, but not by N-ethylmaleimide. This is contrary to the action of these compounds on phosphate-activated glutaminase, demonstrating that the enzyme exerts no effect on the transport [5]. Here we report that the transporter is largely stereospecific, similar to what is found for rat kidney [9] and liver mitochondria [10]. Furthermore, it is saturable, and the uptake of [$^3$H]glutamine shows *cis*-inhibition with *L*-glutamine (not shown), which is also observed for rat submitochondrial particles (SMP) [4].

An electroneutral uniport mechanism has been proposed for glutamine transport in rat renal [3] and liver [11] mitochondria, but in contrast to that in renal mitochondria, the uptake in liver mitochondria was reported to be energy-requiring and driven by the ΔpH [11]. However, our finding that inhibitors of respiration and ionophores inhibit glutamine-induced swelling, suggests that renal mitochondrial glutamine uptake has an energy-requiring component. On the other hand, the concentration gradient across the inner mitochondrial membrane as caused by glutamine consumption by respiration, probably following transamination [1], may promote inward diffusion of glutamine. Since FCCP and nigericin with $K^+$ added inhibit markedly at both pH 6.5 and 7.4, it is indicated that ΔpH also regulates renal glutamine transport, probably by a $H^+$ symport mechanism. In accord with this, raising the pH from 6.5 to 8.0 exerts a pronounced inhibition of the swelling, which contrasts with the finding in experiments with liver mitochondria [11]. In addition, $\Delta\psi$ regulates the uptake since valinomycin inhibits swelling in the presence of $K^+$. It should be noted that valinomycin also exerts a significant, although less pronounced inhibitory effect in nonrespiring mitochondria. With nigericin present and no $K^+$ added, the mitochondria are freely permeable to $H^+$ ions [12]. Under this condition a pH effect still exists, but it is less pronounced. This demonstrates that pig renal glutamine uptake is regulated also by matrix pH, similar to the findings with rat renal SMP [13].

The $K^+$ stimulation of glutamine-induced swelling is intriguing, because the inner mitochondrial membrane has limited permeability to $K^+$. However, a possible explanation is that $K^+$ activates a glutamine transporter, effective on the cytosolic side. Another explanation is that there exists a $K^+$-glutamine symporter. Furthermore, it is of interest that $\Delta\psi$ created in respiring mitochondria can drive the uptake of $K^+$ by a postulated ATP-sensitive channel [14–16]. This uptake can, together with the permeant acetic acid, cause swelling [16]. Thus, it is also a possibility that glutamine, a substrate for respiration, may serve as the permeant compound.

In conclusion, renal mitochondrial transport appears to be complex and may occur by electroneutral as well as by electrophoretic mechanisms.

## References


1 Cooper AJL: Glutamine aminotransferases and ω-amidases; in Kvamme E (ed): Glutamine and Glutamate in Mammals. Boca Raton, CRC Press, 1988, vol 1, pp 33–52.

2 Kovacevic Z, McGivan JD: Mitochondrial metabolism of glutamine and glutamate and its physiological significance. Physiol Rev 1983;63:547–605.

3 Schoolwerth AC, LaNoue KF: Transport of metabolic substrates in renal mitochondria. Annu Rev Physiol 1985;47:143–171.

4 Sastrasinh S, Sastrasinh M: Glutamine transport in submitochondrial particles. Am J Physiol 1989;257:F1050–F1058.

5 Kvamme E, Torgner IAa, Roberg B: Evidence indicating that pig renal phosphate-activated glutaminase has a functionally predominant external localization in the inner mitochondrial membrane. J Biol Chem 1991;266:13185–13192.

6 Kvamme E, Roberg B, Torgner IAa: Effects of mitochondrial swelling and calcium on phosphate-activated glutaminase in pig renal mitochondria. Eur J Biochem 1991;197:675–680.
7 Joseph SK, Meijer AJ: The inhibitory effects of sulphydryl reagents on the transport and hydrolysis of glutamine in rat-liver mitochondria. Eur J Biochem 1981;119:523–529.
8 Lowry OH, Rosebrough NJ, Farr AL, Randall RJ: Protein measurement with the Folin phenol reagent. J Biol Chem 1951;193:265–275.
9 Brosnan JT, Hall B: The transport and metabolism of glutamine by kidney-cortex mitochondria from normal and acidotic rats. Biochem J 1977;164:331–337.
10 Kovacevic Z, Bajin K: Kinetics of glutamine efflux from liver mitochondria loaded with the $^{14}C$-labeled substrate. Biochim Biophys Acta 1982;687:291–295.
11 Soboll S, Lenzen C, Rettich D, Gründel S, Ziegler B: Characterization of glutamine uptake in rat liver mitochondria. Eur J Biochem 1991;197:113–117.
12 Pressman BC: Ionophorous antibiotics as models for biological transport. Fed Proc 1968;27:1283–1288.
13 Sastrasinh M, Sastrasinh S: Effect of acute pH change on mitochondrial glutamine transport. Am J Physiol 1990;259:F863–F866.
14 Halestrap AP, Davidson AM: Inhibition of $Ca^{2+}$-induced large-amplitude swelling of liver and heart mitochondria by cyclosporin is probably caused by the inhibitor binding to mitochondrial-matrix peptidyl-prolyl *cis-trans* isomerase and preventing it interacting with the adenine nucleotide translocase. Biochem J 1990;268:153–160.
15 Bernardi P, Angrilli A, Ambrosin V, Azzone GF: Activation of latent $K^+$ uniport in mitochondria treated with the ionophore A23187. J Biol Chem 1989;264:18902–18906.
16 Beavis AD, Lu Y, Garlid KD: On the regulation of $K^+$ uniport in intact mitochondria by adenine nucleotides and nucleotide analogs. J Biol Chem 1993;268:997–1004.

Prof. Dr. E. Kvamme, Neurochemical Laboratory, Preclinical Medicine,
University of Oslo, PO Box 1115 Blindern, N–0317 Oslo (Norway)

Tizianello A, Baverel G, Endou H, Schoolwerth AC, O'Donovan DJ (eds):
Renal Ammoniagenesis and Interorgan Cooperation in Acid-Base Homeostasis.
Contrib Nephrol. Basel, Karger, 1994, vol 110, pp 13–19

# Effect of α-Ketoglutarate on Ammonia Production in Renal Mitochondria Incubated with a Bicarbonate Buffer[1]

*Sithiporn Sastrasinh, Malinee Sastrasinh*

DVA Medical Center, East Orange, N. J.; University of Medicine and Dentistry, New Jersey Medical School, Newark, N. J., USA

In a deviation from events during acute acidosis in vivo, many studies reported that ammonia production in isolated renal mitochondria decreased in acidic pH [1]. Schoolwerth and LaNoue [2] found that the inhibition of ammonia production was reversed when they added α-ketoglutarate (α-KG) to the incubation buffer. They suggested that the addition of α-KG supplied the substrate for α-KG dehydrogenase flux. Increased α-KG dehydrogenase flux enhanced mitochondrial respiration and $H^+$ efflux which then led to a relatively higher mitochondrial matrix pH ($pH_m$). The maintenance of $pH_m$ and mitochondrial pH gradient (ΔpH) at a relatively high level resulted in higher rate of glutamate dehydrogenase flux and ammonia production. The study of Schoolwerth and LaNoue [2] was done in a nonbicarbonate buffer. There have been many studies of mitochondrial ammoniagenesis in bicarbonate buffer. The purpose of this study is to test the effect of α-KG on ammonia production in renal mitochondria that were incubated with a bicarbonate buffer at an acidic pH.

## Materials and Methods

*Isolation of Mitochondria and Measurement of Ammonia Production*

We used differential centrifugation [3, 4] to isolate mitochondria from renal cortices of male Sprague-Dawley rats (Camm Research, Wayne, N. J., USA) with 250–350 g body weight. We incubated these mitochondria with 1 m*M* *L*-glutamine, with or without 0.5 m*M*

[1] This study was supported by the Research Service of the Department of Veterans Affairs. We thank Gale McArthur for her secretarial assistance.

α-KG, in buffers with two different $HCO_3$ concentrations. Normal $HCO_3$ buffer contained (in m*M*) 100 KCl, 25 $KHCO_3$, 3 $MgSO_4$, 4 ATP, 10 $NaH_2PO_4/Na_2HPO_4$ and 0.8 ethylenediaminetetraacetic acid (EDTA). Low $HCO_3$ buffer contained 115 KCl, 10 $KHCO_3$ and same concentrations of other ingredients. Both buffers were equilibrated with 95% $O_2$–5% $CO_2$. We incubated mitochondria in these buffers at 37 °C for 8 min. At the end of incubation we added 70% perchloric acid (46 $\mu l \cdot mg^{-1}$ protein) to the mitochondrial suspension. The concentration of ammonia in neutralized perchloric acid extract of the buffer was measured with colorimetric method [5].

*Measurement of $pH_m$*

We determined $pH_m$ from the fluorescent ratio of pH-sensitive dye 2′,7′-bis(carboxyethyl)-5(6)-carboxyfluorescein (BCECF). We initially loaded the acetoxymethyl ester of the dye (BCECF-AM) into mitochondria. Mitochondrial esterase hydrolyzes BCECF-AM to form free BCECF. The loading of BCECF-AM into mitochondria followed two previously described methods [6, 7], with minor modifications. We added mitochondria to a loading buffer that contained 0.27 *M* sucrose, 20 m*M* Tris-Hepes and 5 μg/ml of BCECF-AM (Molecular Probes, Inc., Eugene, Oreg., USA) at 25 °C. Mitochondria were left in the loading buffer for 20 min before they were diluted sevenfold with ice-cold sucrose/Tris-Hepes buffer and washed twice with the same buffer. We kept these mitochondria in cold sucrose/Tris-Hepes buffer until the time of $pH_m$ measurement. To measure $pH_m$, we added 20 μl mitochondrial suspension to 2 ml incubation buffer at 37 °C. Mitochondria were continuously stirred in temperature-controlled cuvette and the fluorescent ratio measured at 500/450 nm excitation and 530 nm emission wavelengths (Perkin-Elmer spectrofluorometer model LS 50).

We also determined the leakage of BCECF from mitochondria over time and compared this to the total amount of BCECF loaded into mitochondria. Mitochondria were separated from the buffer with microcentrifuge (10,000 rpm × 1 min) at regular time intervals and the amount of BCECF in the buffer measured from its fluorescence at 450 nm excitation and 530 nm emission wavelengths. We measured total amount of BCECF loaded into mitochondria after we exposed them to 0.03% Triton X-100 for 1 min.

*Calculation and Statistical Analysis*

We calculated ammonia production rate from the amount of accumulated ammonia in the buffer and expressed it as micromole per minute per milligram protein. We used a modified biuret reaction with bicinchoninic acid (Pierce, Rockford, Ill., USA) to measure protein content of the sample. We calculated $pH_m$ from pH calibration plot, obtained with the nigericin method [8]. For pH calibration plot, 20 μl mitochondria was added to buffers at different pH. These buffers contained 100 m*M* KCl, 100 m*M* sucrose, 20 m*M* Tris-Hepes and 10 μ*M* nigericin, pH 6.50–7.75. We measured the fluorescent ratio after mitochondria were continuously stirred in the calibration buffer for 5 min. Before any calculation the fluorescent ratio of BCECF was corrected for background fluorescent, obtained in mitochondria that were not loaded with the dye. We defined ΔpH as the difference between $pH_m$ and buffer pH ($pH_o$).

All results were reported as means ± SE. We used ANOVA to test the simple effects of normal and low $HCO_3$, and of α-KG on ΔpH and mitochondrial ammoniagenesis.

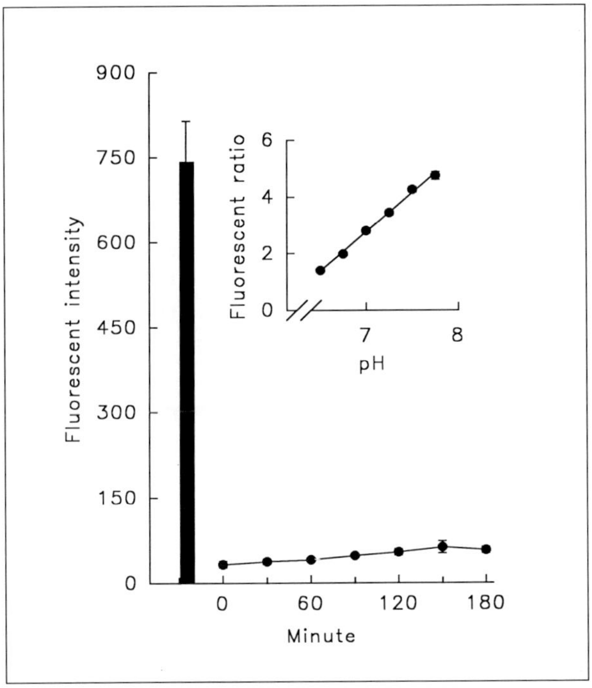


*Fig. 1.* Leakage of BCECF from mitochondria and pH calibration curve. Bar graph shows total amount of free BCECF inside mitochondria. We measured the amount of BCECF released from mitochondria after they were exposed to 0.03% Triton X-100 for 1 min, immediately after BCECF loading. We determined BCECF leakage from mitochondria that were not treated with Triton X-100. The incubation buffer was separated from mitochondria at time points indicated in the graph. The inset shows pH calibration curve of BCECF in this study ($y = 2.82x–17.15$; $r^2 = 0.996$). The fluorescent intensity and fluorescent ratio in this figure are in arbitrary units. The graph shows means ± SE from 5 experiments for BCECF leakage and another 5 experiments for pH calibration plot. The SE are not shown when they are smaller than the graphic representation for the means.

## Results

*Measurement of $pH_m$ with BCECF*

Figure 1 shows the leakage of BCECF from mitochondria and the pH calibration plot. Most of free BCECF remained inside mitochondria after its release from BCECF-AM. Immediately after BCECF loading and resuspension of mitochondria, a small amount (4.3% of total) of BCECF was detected in the buffer. BCECF leakage slightly increased over time (6.6, 7.2 and 7.8% at

*Table 1.* Effect of α-KG and low $HCO_3$ buffer on ammonia production[a]

| Incubation buffer | $NH_3$ production, μmol · min$^{-1}$ · mg$^{-1}$ | | | | | |
|---|---|---|---|---|---|---|
| | 0–4 min | | | 4–8 min | | |
| | normal $HCO_3$ | low $HCO_3$ | p[b] | normal $HCO_3$ | low $HCO_3$ | p[b] |
| (–) α-KG | 0.06 ± 0.005 | 0.05 ± 0.004 | 0.054 | 0.07 ± 0.008 | 0.07 ± 0.008 | 0.303 |
| (+) α-KG | 0.06 ± 0.003 | 0.04 ± 0.003 | 0.008 | 0.07 ± 0.006 | 0.04 ± 0.002 | 0.002 |
| p[c] | 0.422 | 0.098 | | 0.320 | 0.002 | |

[a] Results are means ± SE from 5 experiments.
[b] p values for comparisons between normal and low $HCO_3$ groups.
[c] p values for comparisons between mitochondria incubated either with or without 0.5 m*M* α-KG.

90, 120 and 180 min). These results compared favorably with findings in rat heart mitochondria which had 15% BCECF leakage at the beginning of the experiment [9]. The relationship between $pH_m$ and the fluorescent ratio of BCECF was linear between pH 6.5 and 7.75.

*Effect of α-KG on Mitochondrial Ammoniagenesis*

Table 1 summarizes the data on ammonia production in mitochondria incubated with different buffers. In the absence of α-KG there was no significant difference in the rate of mitochondrial ammonia production whether they were incubated in normal or low $HCO_3$ buffer. When 0.5 m*M* α-KG was included in the buffer it did not enhance ammonia production in low $HCO_3$ buffer. On the contrary, in the presence of α-KG, mitochondrial ammonia production in low $HCO_3$ buffer was significantly lower than in normal $HCO_3$ buffer. Our analysis also failed to reveal any influence of α-KG on mitochondrial ammonia production either in normal or low $HCO_3$ buffer. The only exception was in low $HCO_3$ buffer during the final 4 min of incubation when ammonia production in buffer with α-KG was about 30% less ($p = 0.002$) than in the buffer without α-KG.

*Effect of α-KG on $pH_m$ and ΔpH*

To seek the explanation for the failure of α-KG to enhance ammonia production in our low $HCO_3$ buffer, we determined $pH_m$ and ΔpH of our mitochondrial preparation in different buffers. Without α-KG the $pH_o$ was

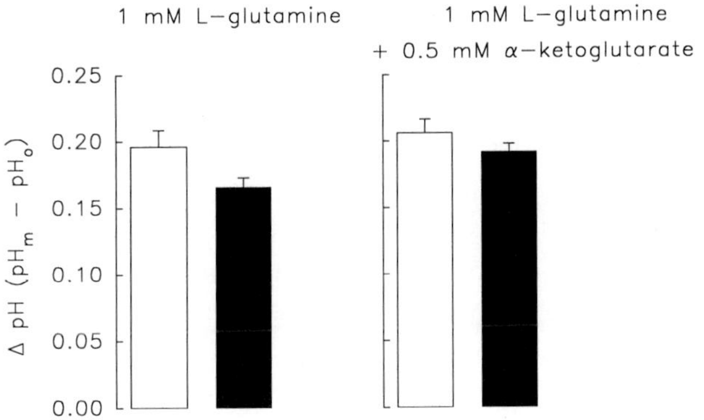


*Fig. 2.* Effect of α-KG on mitochondrial pH gradient. Bar graphs show the difference between mitochondrial matrix pH and buffer pH in mitochondria incubated with different buffers (see details in text). Open bars show data from experiments with normal $HCO_3$ buffer. Solid bars show data from experiments with low $HCO_3$ buffer. Each bar graph represents mean ± SE from 5 experiments.

7.29 ± 0.01 in normal $HCO_3$ buffer and 6.75 ± 0.01 in low $HCO_3$ buffer. With 0.5 m*M* α-Kg the $pH_o$ was 7.27 ± 0.01 in normal $HCO_3$ buffer and 6.72 ± 0.01 in low $HCO_3$ buffer. Figure 2 shows ΔpH in these mitochondria. We did not find any effect of α-KG on ΔpH. In the presence of α-KG, ΔpH in mitochondria incubated with normal and low $HCO_3$ buffers was similar ($p = 0.300$, $n = 5$). Furthermore, ΔpH in mitochondria incubated in low $HCO_3$ buffer with α-KG was not significantly different from ΔpH in mitochondria incubated in low $HCO_3$ buffer, but without α-KG (0.17 ± 0.01 vs. 0.20 ± 0.01 pH unit, $p = 0.064$, $n = 5$).

## Discussion

In this study α-KG did not have any effect on mitochondrial $NH_3$ production at acidic pH. In fact, ammonia production was suppressed further in low $HCO_3$ buffer that contained α-KG. In the absence of α-KG, there was no significant difference in ammonia production whether mitochondria were incubated with normal or low $HCO_3$ buffers. The presence of α-KG markedly suppressed mitochondrial ammonia production in low $HCO_3$ buffer. Mitochondrial ammonia production in low $HCO_3$ buffer with α-KG was 30–40% lower than in normal $HCO_3$ buffer. To explain these findings we focussed our attention on changes in $pH_m$ and ΔpH. In the study of Schoolwerth and LaNoue [2] the addition of α-KG enhanced α-KG dehydrogenase flux, mitochondrial respiration and $H^+$ efflux from mitochondria. These events led to higher $pH_m$ and ΔpH and a higher rate of ammoniagenesis. α-KG failed to induce these changes in our mitochondrial preparation. The method that Schoolwerth and LaNoue [2] used to isolate renal mitochondria differed from ours. But, it was unlikely that difference in isolation techniques was responsible for our failure to elicit any effect of α-KG on mitochondrial ammoniagenesis. Mitochondria isolated according to the techniques that we used have been successfully used to study mitochondrial respiration [4]. We believe that a more likely explanation was the difference in the composition of the incubation buffers. There were two major differences between the buffers used in the two studies. Schoolwerth and LaNoue [2] used a nonbicarbonate buffer that also contained glucose, hexokinase and ADP. We used a bicarbonate buffer that did not have glucose, hexokinase or ADP. The use of bicarbonate buffer has been shown to lower $pH_m$ in rabbit and rat renal cortical mitochondria [10, 11]. It is doubtful, however, that our findings were the result of using a bicarbonate buffer system. In this study the use of either normal or low $HCO_3$ concentration by itself did not have any measurable effect on $pH_m$ and ΔpH (fig. 2). A more plausible explanation for a relatively low $pH_m$ and ΔpH of mitochondria in this study was the use of ATP and the lack of ADP and ADP regenerating system such as glucose/hexokinase in our buffers. In the absence of an ADP regenerating system accumulated ATP will eventually reduce the rate of electron transport through the mitochondrial electron transport chain and the rate of $H^+$ efflux from mitochondria. Additionally, high concentration of ATP will have an inhibitory effect on α-KG dehydrogenase [12] which may further depress mitochondrial respiration. In a buffer with ADP regenerating system continuous ATP hydrolysis will prevent its accumulation. ADP will enter into mitochondria, through the ATP/ADP carrier, undergo oxidative phosphorylation, with continuous cycling of $H^+$ out and back into mitochondria. This will result in a constant flow of electrons through the electron transport chain and continuous $H^+$ efflux from mitochondria.

## References


1 Tannen RL, Sastrasinh S. Response of ammonia metabolism to acute acidosis. Kidney Int 1984;25:1–10.

2 Schoolwerth AC, LaNoue KF: Control of ammoniagenesis by α-ketoglutarate in rat kidney mitochondria. Am J Physiol 1983;244:F399–F408.

3 Sastrasinh S, Sastrasinh M: Glutamine transport in submitochondrial particles. Am J Physiol 1989;257:F1050–F1058.
4 Weinberg JM, Harding PG, Humes HD: Mitochondrial bioenergetics during the initiation of mercuric chloride-induced renal injury. I. Direct effect of in vitro mercuric chloride on renal mitochondrial function. J Biol Chem 1982;257:60–67.
5 McCullough H: The determination of ammonia in whole blood by a direct colorimetric method. Clin Chim Acta 1967;17:297–304.
6 Brierley GP, Davis MH, Cragoe EJ Jr, Jung DW: Kinetic properties of the $Na^+/H^+$ antiport of heart mitochondria. Biochemistry 1989;28:4347–4354.
7 Kapus A, Lukács GL, Cragoe EJ Jr, Ligeti E, Fonyó A: Characterization of the mitochondrial $Na^+$-$H^+$ exchange. The effect of amiloride analogues. Biochim Biophys Acta 1988;944:383–390.
8 Thomas JA, Buschbaum RN, Zimniak A, Racker E: Intracellular pH measurements in Ehrlich ascites tumor cells utilizing spectroscopic probes generated in situ. Biochemistry 1979;18:2210–2218.
9 Kapus A, Ligeti E, Fonyó A: $Na^+/H^+$ exchange in mitochondria as monitored by BCECF fluorescence. FEBS Lett 1989;251:49–52.
10 Simpson DP, Hager SR: Bicarbonate-carbon dioxide buffer system: A determinant of the mitochondrial pH gradient. Am J Physiol 1984;247:F440–F446.
11 Scaduto RC Jr, Schoolwerth AC: Effect of bicarbonate on glutamine and glutamate metabolism by rat kidney cortex mitochondria. Am J Physiol 1985;249:F573–F581.
12 Lehninger AL: Electron transport, oxidative phosphorylation, and regulation of ATP production; in Principles of Biochemistry. New York, Worth, 1982, pp 467–510.

Sithiporn Sastrasinh, MD, Renal-Hypertension Section (111),
DVA Medical Center, 385 Tremont Avenue, East Orange, NJ 07018–1095 (USA)

Tizianello A, Baverel G, Endou H, Schoolwerth AC, O'Donovan DJ (eds):
Renal Ammoniagenesis and Interorgan Cooperation in Acid-Base Homeostasis.
Contrib Nephrol. Basel, Karger, 1994, vol 110, pp 20–24

# ATP Turnover and Intracellular pH Limit Glutamine Utilization in Canine Renal Proximal Tubules[1]

*Michael C. Chobanian, Mark E. Anderson, John L. Markley, Peter C. Brazy*

University of Wisconsin–Madison, Departments of Pediatrics, Medicine, and Biochemistry and the National NMR Facility at Madison, Wisc., USA

*L*-Glutamine is the key substrate for renal proximal tubule ammonia metabolism. Despite knowing for a half a century that glutamine is the key substrate for the ammonia generated by the kidney during acidosis [1], regulation of glutamine utilization is still not fully understood. Recently, a group of Canadian researchers [2] proposed that the upper limit of renal ammonia produced from glutamine is limited by ATP turnover. This theory is supported by in vivo data from acidotic dogs which correlates the amount of ATP necessary to reabsorb sodium (indirectly determined by oxygen consumption) with the ATP derived from *L*-glutamine and lactate metabolism [3]. Additional investigations utilizing suspensions of canine isolated proximal tubules [4] have provided supportive evidence for the theory. When extracellular pH falls to 7.0, all of the ATP turnover in the proximal tubules was found to be dependent upon *L*-glutamine metabolism [4].

Subsequent investigations have refuted the premise that ATP turnover *alone* can account for the upper limit of ammonia production from glutamine. First, glutamine loading both in vivo [5] and in vitro [6] demonstrate an enhancement of ammonia production with no real change in either ATP turnover (indirectly determined by oxygen consumption) or adenine nucleotide precursors. Furthermore, Nagami et al. [7] have provided evidence that by merely increasing perfusion rates in isolated perfused mouse proximal tubules, that

[1] Funded by grants from the National Kidney Foundation of Wisconsin, Department of Veteran Affairs and NIH grant RR02301 (Biomedical Research Technology Program).

ammonia production increased without a change in fluid or sodium reabsorption.

We suggest that factors besides ATP turnover may be important components in limiting the rate of glutamine utilization. ATP turnover refers specifically to that ATP which is being utilized for processes requiring maintenance of cellular ion pumps and nontransport metabolism. ATP turnover calculation assumes that ADP and Pi are not rate limiting. The free energy ($\Delta G_{ATP}$) equation however, by definition, utilizes other key regulators of cellular metabolism including ATP. The free energy equation has been generally expressed as:

$$\Delta G_{ATP} = \Delta G^{\circ}_{ATP} - RT \ln [\alpha_{Mg\,ATP} \cdot \alpha_{Mg\,ADP}{}^{-1} \cdot \alpha_{Pi}{}^{-1} \cdot \alpha_{H+}{}^{-1}].$$

In the equation the change in pH is generally not accounted for but indeed may be a contributing factor. One can rewrite the equation factoring for pH [8]:

$$\Delta G_{ATP} = \Delta G^{\circ}_{ATP} - RT \ln [ATP \cdot ADP^{-1} \cdot Pi^{-1}] - 2.3026\ RT \times pHi.$$

Thus, the free energy equation predicts the amount of ATP necessary to maintain basal cellular processes as a function of ADP, Pi, and pHi [8, 9]. To demonstrate the importance of determining $\Delta G_{ATP}$ instead of ATP turnover, depolarization-induced cellular alkalinization by $BaCl_2$ was used to affect a change in pHi without appreciably affecting ATP utilization (since *L*-glutamine was not rate limiting).

## Materials and Methods

Briefly, proximal tubule segments (PTS) from mongrel dogs were isolated by partial collagenase-digestion and Percoll-density gradient centrifugation as described previously [10]. Aliquots of PTS were suspended in a physiologic pH 7.4 KHB buffer using 10 m*M* *L*-glutamine as the sole substrate (a non-rate-limiting concentration). Six percent perchloric acid–1.0 m*M* EDTA extracts were taken at various time-points for up to 45 min. Ammonia and protein concentrations were determined as before [10]. Oxygen consumption was measured polarigraphically using a Clark oxygen electrode, as described previously [11] and expressed as nanomoles $O_2$ consumed per milligram tissue protein per minute. ATP levels were determined by the luciferase method [12]. In parallel experiments, PTS were placed in a specially designed $^{31}P$-NMR perfusion probe to determine pHi and relative Pi and ATP contents using 10 m*M* *L*-glutamine as the sole substrate [13]. Data are expressed as mean ± SE.

*Table 1.* Effect of $BaCl_2$ upon rate of ammonia production in isolated PTS

| | Rate of ammonia production | |
|---|---|---|
| | 15 min | 45 min |
| Basal conditions | 15.3 ± 6.8 | 15.9 ± 2.7 |
| 0.5 m*M* $BaCl_2$ | 20.1 ± 0.8 | 11.4 ± 1.7 |
| Percent change | p = NS | ↓39.5 ± 2.2* |

Rates of $NH_3$ production expressed as μmol/g protein/min, mean ± SE, for 5 experiments.

* $p < 0.05$ paired t test, 0.5 m*M* $BaCl_2$ vs. basal conditions.

*Table 2.* Effect of barium chloride upon Pi and organic phosphates

| | PME | Pi | β-NTP | pHi |
|---|---|---|---|---|
| Basal conditions | | | | |
| 0.4 m*M* ECP, pHo 7.4 | 2.3 ± 0.83 | 2.4 ± 0.79 | 2.92 ± 1.8 | 7.437 ± 0.04 |
| 0.5 m*M* $BaCl_2$ (4 min) | 3.11 ± 1.9 | 3.70 ± 0.72 | 3.62 ± 2.5 | 7.461 ± 0.04 |
| Percent change | p = NS | ↑26.5 ± 11.2* | p = NS | ↑32.0 ± 0.04 |

Phosphate content given as nmol/mg protein, mean ± SE, for 3 experiments.
* $p < 0.05$ paired t test, 0.5 m*M* $BaCl_2$ vs. basal conditions.

## Results

Upon the addition of 0.5 m*M* $BaCl_2$, rates of ammonia production fell significantly by 45 min when compared to control PTS (table 1). Simultaneous, oxygen consumption rates and ATP contents were unaffected throughout the entire 45-min incubation by the presence or absence of 0.5 m*M* $BaCl_2$ (25.4 ± 0.54 vs. 26.9 ± 0.95 nmol $O_2$ consumed/min/mg protein and 3.62 ± 2.5 vs. 2.92 ± 1.8 nmol/mg protein respectively). In parallel $^{31}$P-NMR experiments, addition of $BaCl_2$ was associated with a significant increase in both Pi and pHi by 4 min (table 2). Intracellular pH continued to rise for up to 15 min (data not shown) to 7.502. Neither β-NTP nor phosphomonoester content was affected for up to 15 min (data not shown). These results demonstrate a clear change in free energy based primarily upon a membrane-depolarized change in pHi or by another ATP-independent mechanism.

## Discussion

Theoretically, all metabolic pathways are regulated by three mechanisms: (1) availability and interaction of substrates; (2) enzyme kinetics and cofactors (i.e. ions, cell signaling, etc.), and (3) bioenergetics of the cell. Regulation of renal ammonia metabolism is a consequence of interactions between each of these mechanisms. When perturbations are introduced, ammoniagenesis may be modified by one or all of the mechanisms in concert. Our results demonstrate a change in ammoniagenesis when pHi is modified by membrane depolarization alone without a significant alteration in the phosphorylation potential or ATP turnover. Although ATP turnover was not determined directly and the results represent acute rather than chronic changes, the findings provide a clear exception to the ATP turnover hypothesis.

Hormonal regulation of renal ammoniagenesis may also provide additional examples of the independence of ATP turnover in the regulation of glutamine utilization. The present authors [14] and others [15] have shown in both canines and rodents that addition of insulin stimulates ammonia production while inhibiting glucose production in vitro using non-rate-limiting concentrations of glutamine. Insulin has also been shown to activate the luminal $Na^+$-$H^+$ antiporter [15] without significantly affecting oxygen consumption rates (data not shown). Theoretically, insulin may be affecting the depolarization of the renal basolateral membrane as has been demonstrated in skeletal muscle [8]. Despite insulin-induced activation of $Na^+$-$H^+$ exchange, insulin may simply enhance glutamine transport as we have previously described [14] without affecting oxygen consumption rates (data not shown), thereby increasing ammonia production.

Our experiments utilizing $BaCl_2$ in canine PTS correlate well with the findings of Siebens and Boron [16] who demonstrated an alkalinization of isolated perfused salamander proximal tubules following introduction of 1.0 m*M* $BaCl_2$ to both the bath and lumen. Hamm and co-workers [17] recently failed to demonstrate a negative effect upon in vivo ammonia production in isolated microperfused rat proximal tubules using 10 m*M* $BaCl_2$ in the perfusate. They were able to provide a negative effect of $BaCl_2$ upon ammonium entry and loss only when EIPA (amiloride analogue) was added to the perfusate bath. Although ammonia production was not measured directly, their findings are difficult to interpret in light of those presented (tables 1, 2) utilizing only 0.5 m*M* $BaCl_2$ except that barium chloride had no access to the basolateral side in the in vivo experiments.

Recently, Vinay and co-workers [18] have demonstrated the presence of an apical $H^+$-ATPase pump in canine, hamster, and rabbit proximal $S_1$-$S_2$ tubules. Under basal physiologic conditions, the $H^+$ pump in canine PTS accounted for nearly 30–45% of the energy demands of the cell as determined by in vitro oxygen consumption rates [7]. Fluctuations in cellular pH alone affected rates of $H^+$ extrusion while the pump remained independent of glycolysis. Thus, as initially presented, the interrelationships of all the regulators of ammonia metabolism in renal proximal tubules including pHi, ATP turnover, etc., may at anytime affect glutamine utilization alone or in concert.

## References


1. Van Slyke DD, Phillips RA, Hamilton PB, Archibald RM, Futcher PH, Miller A: Glutamine as source material of urinary ammonia. J Biol Chem 1943;150:481–483.
2. Brosnan JT, Lowry M, Vinay P, Gougoux A, Halperin M: Renal ammonium production – une vue canadienne. Can J Physiol Pharmacol 1987;65:489–498.
3. Halperin ML, Vinay P, Gougoux A, Pichette C, Jungas RL: Regulation of the maximal rate of renal ammoniagenesis in the acidotic dog. Am J Physiol 1985;248:F607–F615.
4. Manillier C, Vinay P, Lalonde L, Gougoux A: ATP turnover and renal response of dog tubules to pH changes in vitro. Am J Physiol 1986;251:F919–F932.
5. Gougoux A, Vinay P, Halperin ML: Regulation of renal ammoniagenesis in the dog with chronic metabolic acidosis: Effect of a glutamine load. Am J Physiol 1985;249:F745–F752.
6. Chobanian MC, Julin CM, Molteni KH, Brazy PC: Growth hormone regulates ammoniagenesis in canine renal proximal tubule segments. Am J Physiol 1992;262:F878–F884.
7. Nagami GT, Sonu SM, Kurokawa K: Ammonia production by mouse proximal tubules perfused in vitro. Effect of metabolic acidosis. J Clin Invest 1986;78:124–129.
8. Moore RD, Fidelman ML, Hansen JC, Otis JN: The role of intracellular pH in insulin action; in Intracellular pH: Its Measurement, Regulation, and Utilization in Cellular Functions. New York, Liss, 1982, pp 385–416.
9. Veech RL, Lawson JWR, Cornell NW, Krebs HA: Cytosolic phosphorylation potential. J Biol Chem 1979;254:6538–6547.
10. Chobanian MC, Hammerman MR: Phorbol esters inhibit ammoniagenesis and gluconeogenesis in proximal tubular segments. Am J Physiol 1987;252:F1073–F1079.
11. Brazy PC, Gullans SR, Mandel LT, Dennis VW: Metabolic requirement for inorganic phosphate by the rabbit proximal tubule: Evidence for a Crabtree effect. J Clin Invest 1982;70:53–62.
12. Stanley PE, Williams SG: Use of liquid scintillation spectrometer for determining adenosine triphosphate by the luciferase enzyme. Anal Biochem 1969;29:381–392.
13. Anderson ME, Markley JL, Arus C, Chobanian MC: A perifusion loop-gab resonator NMR probe for aerobic cell suspensions. Magn Reson Med 1993;29:563–599.
14. Chobanian MC, Hammerman MR: Insulin stimulates ammoniagenesis in canine renal proximal tubular segments. Am J Physiol 1987;253:F1171–F1177.
15. Gesek F, Mishkind MT, Schoolwerth AC: Hormonal regulation of renal ammonium and glucose production are modulated by $Na^+/H^+$ exchange. Contrib Nephrol. Basel, Karger, 1991, vol 92, pp 141–148.
16. Siebens AW, Boron WF: Depolarization-induced alkalinization in proximal tubules. I. Characteristics and dependence on $Na^+$. Am J Physiol 1989;256:F342–F353.
17. Simon EE, Merli C, Herndon J, Cragoe EJ Jr, Hamm LL: Effects of barium and 5-(N-ethyl-N-isopropyl)amiloride on proximal tubule ammonia transport. Am J Physiol 1992;262:F36–F39.
18. Noël J, Vinay P, Tejedor A, Fleser A, Laprade R: Metabolic cost of bafilomycin-sensitive $H^+$ pump in intact dog, rabbit, and hamster proximal tubule. Am J Physiol 1993;264:F655–F661.



M.C. Chobanian, MD, H4/469 Clinical Science Center, University of Wisconsin School of Medicine, 600 Highland Avenue, Madison, WI 53792–4108 (USA)

Tizianello A, Baverel G, Endou H, Schoolwerth AC, O'Donovan DJ (eds):
Renal Ammoniagenesis and Interorgan Cooperation in Acid-Base Homeostasis.
Contrib Nephrol. Basel, Karger, 1994, vol 110, pp 25–29

# Effect of Acute Acidosis on the Pathways of Ammoniagenesis in LLC-$PK_1$ Cells

*Itzhak Nissim*[a], *Atul Sahai*[b], *Richard L. Tannen*[b]

a Division of Biochemical Development and Molecular Diseases, Children's Hospital of Pennsylvania, University of Pennsylvania School of Medicine, Philadelphia, Pa., USA
b Department of Medicine, University of Southern California School of Medicine, Los Angeles, Calif., USA

There is now considerable evidence that an acute reduction in pH stimulates renal ammonia production [1]. However, the precise mechanism and pathways of ammonia formation altered by acute acidosis have not been fully clarified. These issues have been addressed using several experimental models including isolated mitochondria, intact renal tissues and cultured renal tubular cells. The results, however, have been conflicting.

Studies of isolated renal cortical mitochondria from the dog initially indicated that a low pH inhibited glutamine deamidation; however, in subsequent studies carried out with a bicarbonate buffer system, ammonia production was stimulated [2, 3]. Since glutamate metabolism was inhibited in these studies, the heightened ammoniagenesis resulted from increased flux through phosphate-dependent glutaminase. When rat renal cortical mitochondria are incubated with glutamine alone, ammonia production is inhibited by a low pH regardless of the buffer system used [3–6]. With the addition of α-ketoglutarate to the medium, a low pH stimulates ammonia production by rat mitochondria entirely as a result of increased flux through glutamate dehydrogenase [7]. Of interest, studies utilizing submitochondrial particles indicate that a low pH can stimulate the mitochondrial glutamine transporter [8]. Thus, studies with isolated mitochondria indicate that a low pH can stimulate ammoniagenesis, but is unclear whether flux through PDG or GLDH is primarily affected.

Studies of intact renal tissue and cultured renal tubular epithelium have also been conflicting. Using $^{15}N$-labelled glutamine, Nissim and co-workers

*Table 1.* Glutamine utilization and $^{15}N$ production: comparison of pH values

| pH | Glutamine utilization nmol/h/mg | $^{15}N$ production, nmol/h/mg | |
|---|---|---|---|
| | | $NH_3$ | alanine |
| *5-$^{15}N$ glutamine* | | | |
| 7.0 | 807±51* | 149±9 | |
| 7.4 | 475±88 | 140±6 | |
| 7.6 | 390±69* | 147±7 | |
| *2-$^{15}N$ glutamine* | | | |
| 7.0 | 754±46* | 58±8* | 159±10* |
| 7.4 | 497±56 | 22±3 | 110±11 |
| 7.6 | 416±77* | 13±1* | 96±11* |

* Significantly different from pH 7.4.

[9, 10] found that a low pH stimulated flux through GLDH by both cultured rat and human proximal tubular epithelium. By contrast, Bogusky and Dietrich [11] found that a low pH stimulated ammonia production by the isolated perfused rat kidney entirely as a result of increased flux through PDG without any change in flux through GLDH.

Our laboratory had found that LLC-$PK_1$ cells, a pig kidney-derived epithelial cell line with characteristics of proximal tubule epithelium, exhibit characteristics of ammonia metabolism comparable to intact renal tissue [12]. These cells increase ammonia production in response to acute acidosis and analysis based on net production of metabolites suggested that this results from increased flux through PDG and not GLDH [13].

Therefore we utilized $^{15}N$-labelled glutamine to clarify the precise alterations in glutamine metabolism produced when LLC-$PK_1$ cells are exposed to a low pH. We also tested the response to different degrees of acidosis.

## Methods

Stock cultures of chronically rocked LLC-$PK_1$ cells were subcultured in 60-mm dishes in minimal essential medium containing 10% fetal bovine serum and 2 m*M* glutamine. When the cultures reached 75–85% confluency, the cells were washed with phosphate-buffered saline and incubated for 1 h in Krebs-Henseleit buffer (KHB) at 37 °C in a 5% $CO_2$/95% air environment. The medium was supplemented with 5 m*M* glucose and 1 m*M* $^{15}N$-glutamine labelled at either the 5-N or 2-N position. The pH of the medium was adjusted to 7.6, 7.4, 7.0 or 6.8, by modifying the bicarbonate concentration.

In the initial experiments (shown in table 1), the reaction was terminated by adding 100 µl of 1 *N* HCl to the media, while in the subsequent experiments (shown in table 2)

*Table 2.* $^{15}N$ production: comparison of pH values

| pH | $^{15}N$ production, nmol/h/mg | |
|---|---|---|
| | $NH_3$ | alanine |
| *5-$^{15}N$ glutamine* | | |
| 7.4 | 170±10 | |
| 7.0 | 195±11 | |
| 6.8 | 224±9* | |
| *2-$^{15}N$ glutamine* | | |
| 7.4 | 25±2* | 99±4* |
| 7.0 | 49±4 | 164±11 |
| 6.8 | 94±6* | 180±9 |

* Significantly different from pH 7.0.

the KHB was removed and mixed with a 6% perchloric acid to terminate the reaction. Appropriate controls were treated similarly and the experimental data were corrected accordingly.

Measurement of cell protein content and the analysis of $^{15}N$ metabolites were carried out as described previously [14, 15]. A value of $p < 0.05$ was considered significant.

## Results

In the initial sequence of studies, shown in table 1, a comparison was made between cells incubated at pH 7.6, 7.4 and 7.0. At pH 7.0, glutamine utilization was stimulated; whereas at pH 7.6 it was inhibited when studies were carried out with either 5 or 2-$^{15}N$ glutamine as substrate.

Production of $^{15}N$-$NH_3$ was not altered by either a low or high pH when 5-$^{15}N$ glutamine was used as substrate. By contrast a low pH stimulated and a high pH decreased $^{15}N$-$NH_3$ production in studies utilizing 2-$^{15}N$ glutamine. Thus an increase in pH to 7.6 and a decrease to 7.0 altered ammonia production solely as a result of changes in the rate of glutamine deamination. Glutamine deamidation was not modified by acute pH changes within this range. $^{15}N$ alanine production was enhanced at pH 7.0 and diminished at pH 7.6 in comparison with pH 7.4 when 2-$^{15}N$ glutamine was used as substrate.

In a second set of studies, shown in table 2, a comparison was made between cells incubated at pH 7.4, 7.0 and 6.8. Consistent with the findings in the initial studies, a decrease in pH from 7.4 to 7.0 did not alter the rate of $^{15}N$-$NH_3$ production when 5-$^{15}N$ glutamine was used as substrate. However, a decrease in pH to 6.8 significantly stimulated $^{15}N$-$NH_3$ production from 5-$^{15}N$

glutamine in comparison with both pH 7.0 and 7.4. $^{15}N$-$NH_3$ production with 2-$^{15}N$ glutamine as substrate was increased significantly at pH 7.0 in comparison with pH 7.4, and increased yet further at pH 6.8. $^{15}N$ alanine production was increased at pH 7.0 in comparison with pH 7.4, but in contrast to ammonia production from 2-$^{15}N$ glutamine, there was no further increment at pH 6.8.

## Discussion

Previous studies from our laboratory have demonstrated that adequately oxygenated LLC-$PK_1$ cells exposed to a low media pH exhibit a decrease in intracellular pH and a concurrent increase in ammonia production [13]. This is accompanied by a decrease in the intracellular concentration of both glutamate and α-ketoglutarate, similar to the findings in intact renal tissue, and is consistent with the data indicating that a low pH stimulates α-ketoglutarate conversion to succinate by enhancing the activity of α-ketoglutarate dehydrogenase [7, 16, 17].

At pH 7.0, the increase in ammoniagenesis is accounted for entirely by an increase in glutamine deamination, i.e. production of $^{15}NH_3$ from 2-$^{15}N$ glutamine. This increase in flux through glutamate dehydrogenase presumably is secondary to the decrease in α-ketoglutarate concentration produced by pH activation of α-ketoglutarate dehydrogenase.

As we have discussed in a prior publication, the increase in alanine production apparently reflects an increased rate of glutamine, rather than glutamate transamination [15]. This conclusion is based on a quantitative analysis of the $^{15}N$ products of glutamine metabolism and is supported by a concurrent increase in α-ketoglutaramate accumulation.

When the pH is decreased from 7.0 to 6.8, ammonia production from glutamine deamination is further increased. At this low pH, however, the rate of glutamine deamidation, i.e. production of $^{15}NH_3$ from 5-$^{15}N$ glutamine, is also increased. Thus it appears that a sufficiently severe acute acidotic stimulus can enhance flux through PDG as well as through GLDH.

Theoretically, the increase in $^{15}NH_3$ production from 5-$^{15}N$ glutamine could result from stimulation of deamidation through γ-glutamyltranspeptidase or from the glutamine transamination-ω-amidase pathway. Since we have shown previously that ammonia production by LLC-$PK_1$ cells is not influenced by virtually complete inhibition of γ-glutamyltranspeptidase activity, it seems unlikely that this pathway accounts for the low pH-induced increase in glutamine deamidation [12]. It is not possible to completely exclude a role for the glutamine transamination pathway; however, the absence of a change in alanine production between pH 6.8 and 7.0 suggests that this is a unlikely possibility.

Thus increased flux through PDG most likely accounts for the increased rate of glutamine deamidation at pH 6.8. Whether this is secondary to further deinhibition of PDG, secondary to increased flux through GLDH, and/or from a direct effect of pH on the mitochondrial glutamine transporter, is not resolved by these studies. These observations do suggest, however, that the discrepancies between prior investigations concerning the metabolic site of action of acute acidosis might, in part, be related to differences in the intensity of the acidotic stimulus.

## References


1 Tannen RL: Renal ammonia production and excretion; in Windhager E (ed): Handbook of Renal Physiology, ed 2, sect 8. New York, Oxford University Press, 1992, pp 1017–1059.

2 Simpson DP, Adams W: Glutamine transport and metabolism by mitochondria from dog renal cortex: General properties and response to acidosis and alkalosis. J Biol Chem 1975; 250:8148–8158.

3 Simpson DP: Modulation of glutamine transport and metabolism in mitochondria from dog renal cortex: Influence of pH and bicarbonate. J Biol Chem 1980;255:7123–7128.

4 Tannen RL, Kunnin AS: Effect of pH on ammonia production by renal mitochondria. Am J Physiol 1976;231:1631–1637.

5 Schoolwerth AC, Nazar BL, LaNoue KF: Glutamate dehydrogenase activation and ammonia formation by rat kidney mitochondria. J Biol Chem 1978;253:6177–6183.

6 Kunin AS, Tannen RL: Regulation of glutamine metabolism by renal cortical mitochondria. Am J Physiol 1979;237:F55–F62.

7 Schoolwerth AC, LaNoue KF: Control of ammoniagenesis by alpha-ketoglutarate in rat kidney mitochondria. Am J Physiol 1983;244:F399–F408.

8 Sastrasinh M, Sastrasinh S: Effect of acute pH change on mitochondrial glutamine transport. Am J Physiol 1990;259:F863–F866.

9 Nissim I, States B: Ammoniagenesis by cultured human renal cortical epithelia: Study with $^{15}N$. Am J Physiol 1989;256:F187–F196.

10 Nissim I, States B, Yudkoff M, Segal S: Characterization of amino acid metabolism by cultured rat kidney cells: Study with $^{15}N$ glutamine. Am J Physiol 1987;253:F1243–F1252.

11 Bogusky RT, Dietrich RL: Effect of acute metabolic acidosis on ammonia metabolism in kidney. Am J Physiol 1989;256:F321–F328.

12 Sahai A, Cole LA, Tannen RL: Pathways and regulation of ammoniagenesis by LLC-$PK_1$ cells in cultures. J Lab Clin Med 1989;114:285–293.

13 Sahai A, Laughrey E, Tannen RL: Relationship between intracellular pH and ammonia metabolism in LLC-$PK_1$ cells. Am J Physiol 1990;258:F103–F108.

14 Nissim I, Yudkoff M, Segal S: Metabolism of glutamine and glutamate by rat renal tubules: Study with $^{15}N$ and gas chromatography-mass spectrometry. J Biol Chem 1985;260:13955–13967.

15 Sahai A, Nissim I, Tannen RL: Pathways of acute pH regulation of ammoniagenesis in LLC-$PK_1$ cells: Study with $^{15}N$ glutamine. Am J Physiol 1991;261:F481–F487.

16 Tannen RL, Kunin AS: Effect of pH on metabolism of alpha-ketoglutarate by renal cortical mitochondria. Am J Physiol 1981;240:F120–F126.

17 Lowry M, Ross BD: Activation of oxoglutarate dehydrogenase in the kidney in response to acute acidosis. Biochem J 1980;190:771–780.



Richard L. Tannen, MD, LAC-USC Medical Center, Department of Medicine, IRD 206, 1200 N. State Street, Los Angeles, CA 90033 (USA)

Tizianello A, Baverel G, Endou H, Schoolwerth AC, O'Donovan DJ (eds):
Renal Ammoniagenesis and Interorgan Cooperation in Acid-Base Homeostasis.
Contrib Nephrol. Basel, Karger, 1994, vol 110, pp 30–40

# pH and Hormonal Regulation of Ammoniagenesis in Cultured OK Cells[1]

*Itzhak Nissim, Beatrice States, Ilana Nissim, Zhi-Ping Lin, Marc Yudkoff*

Division of Biochemical Development and Molecular Diseases,
The Children's Hospital of Philadelphia, University of Pennsylvania
School of Medicine, Department of Pediatrics, Philadelphia, Pa., USA

The complexity of renal ammonia metabolism is reflected in part by the biochemical and physiological differences in renal tissue of various species as well as the cellular heterogeneity of the nephron. Thus, a variety of renal tissue preparations from assorted species has been used for the investigation of the mechanisms regulating ammoniagenesis in various acid-base states [1].

Over the past decade, interest has grown in the use of cultured kidney cells as a model with which to elucidate the mechanisms controlling renal ammonia metabolism [2–9]. Cultured cells allow the investigation of metabolic and biochemical regulation in a uniform preparation. However, the cellular diversity within the nephron may limit the utility of cultured renal cells for studies of ammoniagenesis because changes of acid-base homeostasis affect ammonia metabolism primarily in the proximal tubules [10]. A more serious concern is that the metabolic properties of established cell lines may not retain the biochemical and metabolic characteristics of the original tubular cells. Nevertheless, the preparation of homogeneous proximal tubule cell type [8, 11, 12], and the possibility of using either primary cultures or subcultures [13], has increased the application of cultured renal cells as a model system for the study of renal ammonia metabolism.

LLC-$PK_1$ renal epithelial cells have been widely used for studies of renal ammonia metabolism [5–7, 9], but even this system has been the subject of conflicting observations. Numerous studies have shown that rocking and oxygenation are required for induction of pH-modulated ammoniagenesis by

[1] Supported by NIH grant RO1-DK 39348 and in part by Grant-in-Aid No. 9100677 from The American Heart Association.

$LLC\text{-}PK_1$ cells [5–7], but a recent study by Gstraunthaler et al. [9] indicates that these cell lines respond to metabolic acidosis under still culture conditions. Similarly, studies using NRK cells [2], cultured human renal epithelial cells [3], or cultured rat renal papillary cells [14], have shown that glutamine metabolism and ammonia production respond to changing medium pH in still culture.

An additional cell line which has been widely used in studies of renal physiology, biochemistry and transport are epithelial proximal tubules (OK cells) derived from American opossum kidney [8]. Previous studies have demonstrated that OK cells have many characteristics of proximal tubules, including $Na^+$-$H^+$ exchange [15], receptors for parathyroid hormone (PTH) [15, 16] and insulin (INS) [17], as well as $\alpha_2$-receptor agonists [18]. However, only scattered reports employing OK cells to study renal ammonia metabolism have appeared in the literature. In the present study we have explored the use of OK cells as a model system for studying the effect of media pH on glutamine metabolism and ammonia production as well as the hormonal regulation of ammonia metabolism at physiologic pH.

## Materials and Methods

The OK cell line, passage 73, was the generous gift of Dr. Joseph Handler. Experiments were carried out using OK cells between passage 79 and 82. Cells were grown in T-25 flasks with α-Minimal Eagle's medium supplemented with 4 m*M* glutamine plus 3% serum (complete α-MEM). Approximately $1 \times 10^5$ cell suspension/ml medium was added to each T-25 flask. Then, 5 ml of α-MEM was added to each flask. Cells were refed every 24 h until they reached 85–90% confluence, i.e., 6–7 days. At this time the steady-state medium was removed and cultures were washed with PBS. To the flasks containing the cell monolayer we then added 5 ml of Krebs-Henseleit buffer at pH 6.8 (acidosis), 7.4 (control) or 7.6 (alkalosis), supplemented with 1 m*M* of either [2-$^{15}$N]glutamine or [5-$^{15}$N]glutamine. The desired pH was achieved by increasing (pH 7.6) or decreasing (pH 6.8) the bicarbonate concentration in the incubation medium.

A separate series of experiments was carried out at pH 7.4 in the presence of either PTH (1–34, human; ICN Biomedicals No. 152976), insulin (Sigma No. I-1882) or glucagon (Sigma G-201S) at final concentration of $10^{-7}$ *M*.

Cells were incubated for 3 h at 37 °C in 5% $CO_2$/95% atmospheric air. Incubations were stopped by addition of perchloric acid (PCA), final concentration 5% (w/v). After neutralization with KOH, the supernatants were used for determination of metabolite concentration as well as $^{15}$N enrichment using GC-MS as previously described [2, 3].

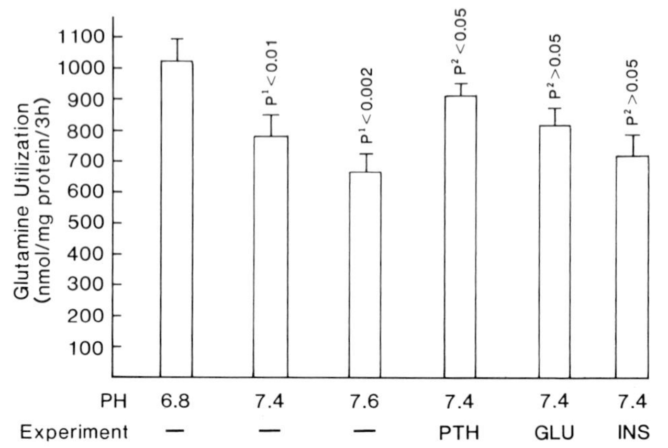


*Fig. 1.* [5-$^{15}$N]glutamine utilization in response to acute alteration of media pH or following addition of PTH, glucagon (GLU) or insulin (INS) to incubation media at pH 7.4. Values of means and error lines are ± SD of 3–5 separate experiments. $p^1$ compared with incubation at pH 6.8 and $p^2$ compared with pH 7.4 without hormone.

## Results

Figure 1 illustrates the utilization of [5-$^{15}$N]glutamine by OK cells following 3 h incubation. The rate of [2-$^{15}$N]glutamine utilization showed little change compared with [5-$^{15}$N]glutamine (data not shown). There was a significantly higher metabolism of glutamine at pH 6.8 compared with either pH 7.4 or 7.6. The addition of insulin or glucagon at pH 7.4 had little effect on glutamine metabolism. However, supplementation of $10^{-7}$ *M* PTH to the incubation medium significantly augmented the utilization of glutamine compared with incubation at pH 7.4.

The primary product of glutamine metabolism is ammonia. Figure 2 demonstrates the formation of $^{15}$N-labeled ammonia from [5-$^{15}$N]glutamine via glutaminase activity. With 1 m*M* [5-$^{15}$N]glutamine as precursor, the enrichment of $^{15}NH_3$ (atom % excess) ranged from 60 to 70% under all experimental conditions, indicating that approximately 60–70% of ammonia-N was derived from 5-N of glutamine at either medium pH or in the presence of hormones. The product of total ammonia formation (nmol/mg protein) times isotopic enrichment reveals a significant increase in the rate of $^{15}NH_3$ production via glutaminase at pH 6.8 compared with pH 7.4 or 7.6. In the presence of glucagon or insulin, $^{15}NH_3$ production increased by ≈20–25% ($p > 0.05$) compared with

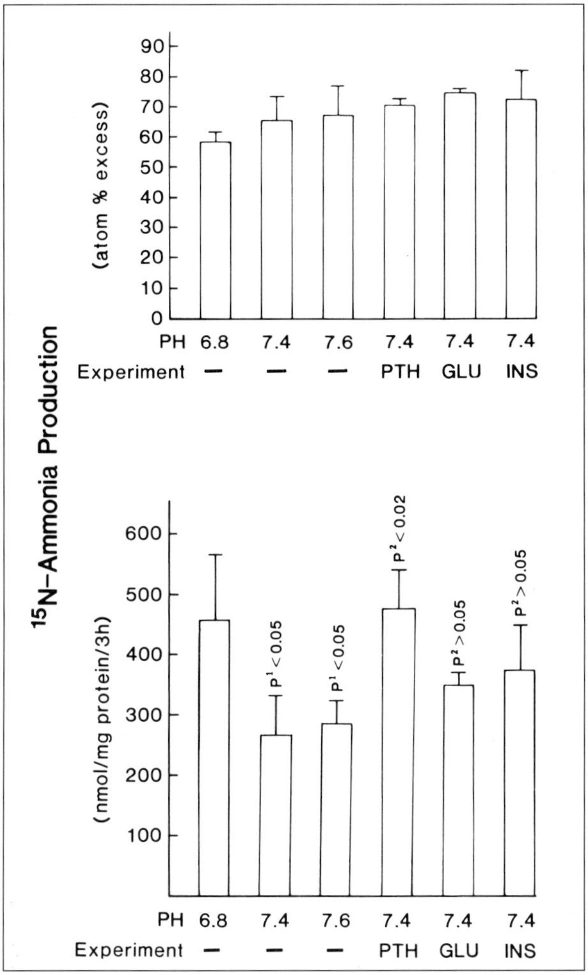


*Fig. 2.* Formation of [$^{15}$N]ammonia enrichment (atom % excess), top, and production, bottom, from 1 m*M* [5-$^{15}$N]glutamine. The production rates are the product of isotopic enrichment (atom % excess) times concentrations (nmol/mg protein) following 3 h incubation. Values are means and error lines are ± SD of 3–5 experiments. $p^1$ compared with incubation at pH 6.8 and $p^2$ compared with pH 7.4 without hormone.

incubation at pH 7.4 without hormone. However, with the addition of PTH to the medium, $^{15}NH_3$ production was significantly increased to a level similar to that obtained at pH 6.8.

Formation of ammonia via the flux through the GLDH pathway is depicted in figure 3. The experiments with [2-$^{15}$N]glutamine demonstrate a sig-

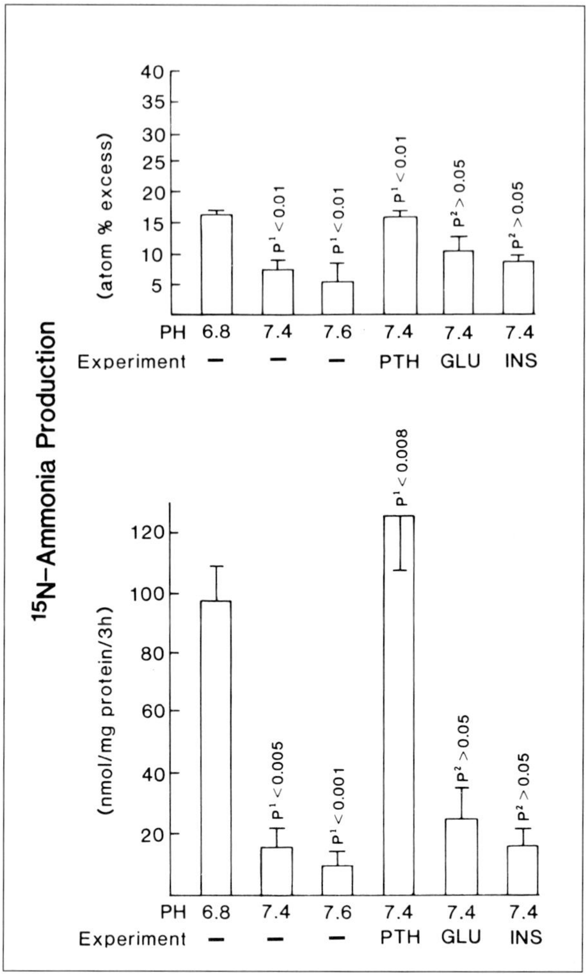


*Fig. 3.* Formation of [$^{15}$N]ammonia from 1 m*M* [2-$^{15}$N]glutamine. Details are as indicated in the legend to figure 2.

nificantly higher $^{15}NH_3$ enrichment (atom % excess) and production (nmol/mg protein) at pH 6.8 compared with incubation at pH 7.4 or 7.6. Only little difference was obtained between pH 7.4 and 7.6. Addition of insulin or glucagon had little effect on the flux through the GLDH reaction. However, PTH enhanced ammonia production from 2-N of glutamine by approximately fourfold ($p < 0.01$) compared with incubation at pH 7.4 without hormone (fig. 3).

An additional product of [2-$^{15}$N]glutamine metabolism is [$^{15}$N]glutamate formed following glutaminase activity. The data of figure 4 illustrate signifi-

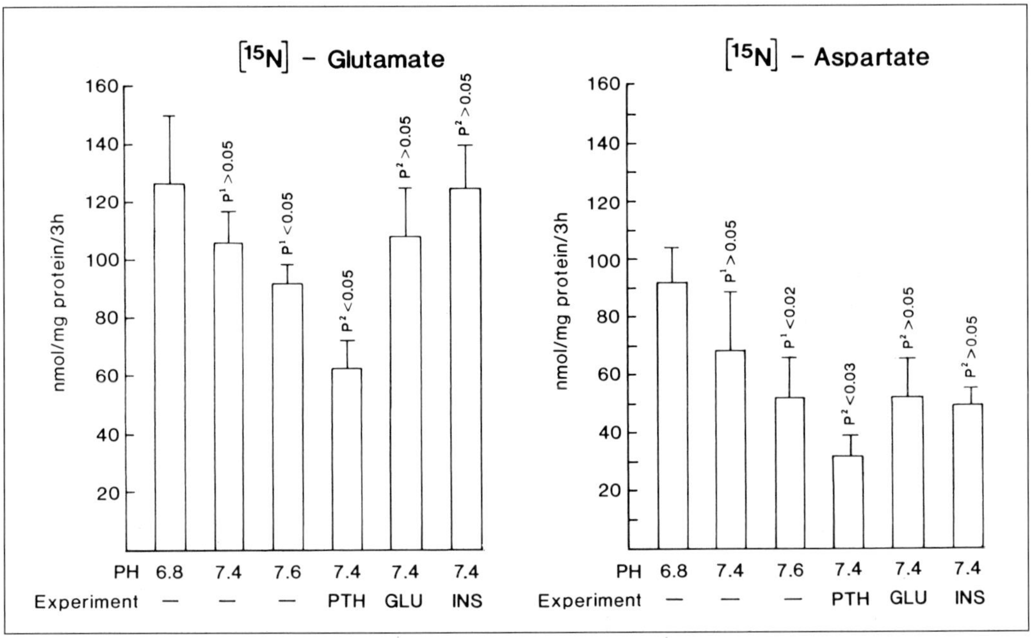


4

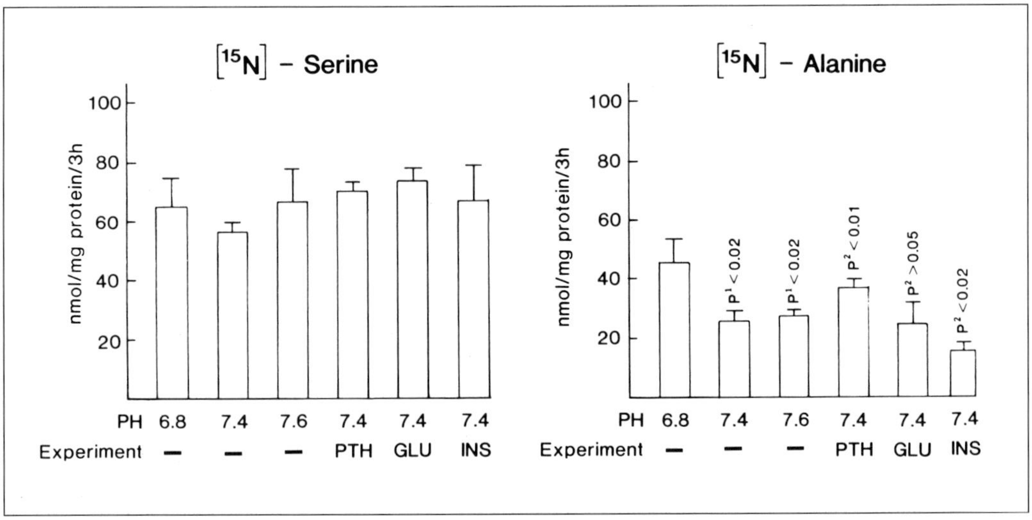


5

*Fig. 4.* Formation of $^{15}N$-labeled glutamate and aspartate following 3 h incubation of OK cells with 1 m*M* [2-$^{15}N$]glutamine. Values are the products of isotopic enrichment times concentrations. Bars are means and error lines are ± SD of 3–5 experiments. $p^1$ compared with incubation at pH 6.8 and $p^2$ compared with pH 7.4 without hormone.

*Fig. 5.* Production of [$^{15}N$]serine and [$^{15}N$]alanine from 1 m*M* [2-$^{15}N$]glutamine. Details are as in the legend to figure 4.

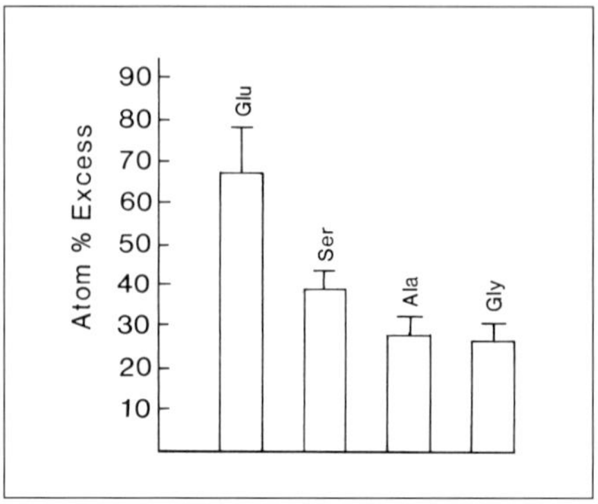


*Fig. 6.* $^{15}$N enrichment in amino acid following 3 h incubation with 1 m*M* [2-$^{15}$N]glutamine. Bars are means of all experiments with various media pH and in the presence of various hormone as indicated in figures 4 and 5. Error lines are ± SD of 20–25 determinations.

cantly higher formation of [$^{15}$N]glutamate at pH 6.8 compared with 7.6 and little difference compared with pH 7.4. Insulin and glucagon have little effect on the production of [$^{15}$N]glutamate; however, PTH markedly diminished the accumulation of glutamate from glutamine (fig. 4). Production of [$^{15}$N]aspartate paralleled that of [$^{15}$N]glutamate, indicating a rapid conversion of glutamate to aspartate and a near equilibrium between these amino acids via GOT reaction. However, contrary to previous studies using LLC-$PK_1$ cells [7] the current data demonstrate a significantly higher $^{15}$N labelling of aspartate (fig. 4) and serine compared with [$^{15}$N]alanine (fig. 5). Furthermore, the isotopic enrichment in serine is significantly higher than alanine, but smaller than glutamate (fig. 6), indicating a precursor-product relationship between glutamate and serine. Serine could have been derived from glutamate via glutamate 3-P-hydroxypyruvate transaminase (3-PHPT). Therefore, the current data suggest a higher flux of glutamate-N via 3-PHPT than through the GPT reaction.

## Discussion

Our purposes in the current study were: (a) to characterize glutamine and ammonia metabolism by OK cells; (b) to evaluate the effect of hormones or media pH on ammoniagenesis, and (c) to determine whether OK cells can be utilized as an alternative model system to study renal N metabolism in vitro. As illustrated in table 1, the results indicate that these cell lines share important metabolic properties with other kidney cell lines previously employed [2–9]. OK cells consume glutamine and respond to the stimulus of acidosis with increased ammonia formation via mitochondrial glutaminase as well as glutamate dehydrogenase. The opposite occurred in response to acute alkalo-

*Table 1.* Glutamine metabolism and ammoniagenetic properties of OK cells compared with other kidney cell lines

| Cell line | Media pH | Glutamine utilization, nmol/mg protein/h | Flux, nmol/mg protein/h | | | | Reference |
|---|---|---|---|---|---|---|---|
| | | | glutaminase | GLDH | GOT | GPT | |
| NRK | 7.4 | 740±255 | 471±51 | 193±23 | 109±15 | 29±2 | 2 |
| | 7.0 | 826±128 | 507±18 | 349±17 | 723±16 | 19±6 | |
| LLC-$PK_1$ | 7.0 | 807±51 | 149±9 | 58±8 | 15±3 | 159±10 | 7 |
| | 7.4 | 475±88 | 140±6 | 22±3 | 20±4 | 110±11 | |
| | 7.6 | 390±69 | 147±7 | 13±1 | 21±6 | 96±11 | |
| OK cells | 6.8 | 344±21 | 151±35 | 32±4 | 30±4 | 15±3 | current study |
| | 7.4 | 262±24 | 89±25 | 11±3 | 22±6 | 8±1 | |
| | 7.6 | 219±22 | 99±16 | 7±2 | 17±4 | 9±1 | |

sis (fig. 1–3, table 1). These observations are in agreement with previously published studies [2, 3], suggesting that even without rocking [4–6], media pH exerted a definite effect on glutamine metabolism and ammonia formation. Furthermore, the current data indicate that PTH mimics the effect of acute acidosis on ammoniagenic pathways.

Previous studies using OK cells have demonstrated that PTH caused a decrease in the rate of $Na^+$-$H^+$ exchange, decreased intracellular pH and increased cAMP [15, 16]. Other studies by Bourke and co-workers [19] have indicated that PTH increased the $NAD^+$/NADH ratio. In the current study, it is possible that PTH stimulated the flux via both the glutaminase and glutamate dehydrogenase pathways as a result of decreased activitiy of the $Na^+$-$H^+$ exchanger and induction of intracellular acidosis. However, an increase in the $NAD^+$/NADH ratio, protein kinase C activity or modulation of intracellular $Ca^{2+}$ may augment flux through the GLDH pathway and, therefore, increase ammonia production from [2-$^{15}$N]glutamine as indicated in figure 3.

The immediate modulator of the GLDH reaction is the mitochondrial $NAD^+$/NADH ratio [1] which PTH increases [11]. In the presence of PTH the formation of $^{15}NH_3$ from [2-$^{15}$N]glutamine via GLDH reaction exceeded that at acute acidosis (fig. 3); whereas, $^{15}NH_3$ production via the glutaminase pathway approximated that at pH 6.8 (fig. 2). However, in the same experiments PTH significantly reduced the formation of [$^{15}$N]glutamate and [$^{15}$N]aspartate (fig. 4). Therefore, the current observation, together with previous studies [15–19], suggests that the effect of PTH on ammonia formation is mainly via the GLDH pathway, presumably due to an increased $NAD^+/NADH$ ratio. This effect is more prominent than that on the glutaminase pathway caused by decreased $Na^+$-$H^+$ exchanger activity and reduced cellular pH. Further studies are necessary to determine the site(s) and the mechanism(s) of PTH action.

Although insulin and glucagon play key roles in the regulation of cellular physiology, metabolism and amino acid transport, the current data indicate little effect of these hormones on glutamine and ammonia metabolism by OK cells. These observations may be explained by the lack of effect of these hormones on $Na^+$-$H^+$ exchange and intracellular pH [18, 19].

$^{15}N$-labeled amino acids presumably were derived from [$^{15}N$]glutamate formed from [2-$^{15}N$]glutamine (fig. 4, 5). These observations demonstrate highly active aminotransferase reactions in the OK cells employed in the current studies. Similar to the LLC-$PK_1$ cell line [7], the formation of [$^{15}N$]alanine was significantly higher at pH 6.8 compared with pH 7.4 or 7.6 (fig. 5). Furthermore, the effect of media pH on the production of [$^{15}N$]alanine paralleled that of $^{15}NH_3$ (fig. 2, 3). [$^{15}N$]alanine could have been derived via GPT from [$^{15}N$]glutamate or via the glutaminase II pathway from [2-$^{15}N$]glutamine. Although in the current studies we were unable to detect production of α-ketoglutaramate as previously indicated [20], we cannot rule out the possibility that some alanine was formed via the glutaminase II pathway.

Contrary to our previous studies using LLC-$PK_1$ [7], the current observations demonstrate a significantly higher accumulation of $^{15}N$-labeled glutamate and aspartate at pH 6.8 compared with pH 7.6 (fig. 4). In the current studies the ratio between $^{15}NH_3$ formation from [2-$^{15}N$] and [5-$^{15}N$]glutamine is 0.2 and 0.07 at pH 6.8 and 7.6, respectively. In our previous studies using LLC-$PK_1$ this ratio was 0.4 and 0.09 at acute acidosis and alkalosis, respectively [7]. Therefore, although the GLDH reaction is sensitive to alteration of media pH, the overall flux through the GLDH pathway is more prominent in LLC-$PK_1$ cells than OK cells. The magnitude of the flux through glutaminase in OK cells is approximate that in LLC-$PK_1$ cells regardless of media pH (table 1). Thus, the accumulation of $^{15}N$-labeled glutamate and aspartate at pH 6.8 perhaps reflects relatively less active mitochondrial GLDH reaction in OK cells compared with LLC-$PK_1$ cells. This may occur due to a lower $NAD^+/NADH$ ratio in the opossum cells.

An observation of special interest is the significantly higher production of [$^{15}N$]serine compared with [$^{15}N$]alanine regardless of the experimental conditions (fig. 5). [$^{15}N$]serine could have been derived from alanine via serine pyruvate transaminase or from [$^{15}N$]glutamate via glutamate-3-P-OH-pyruvate transaminase pathway. The isotopic enrichment in serine is higher than that in alanine but smaller than glutamate (fig. 6). Therefore, a precursor-product relationship exists between glutamate and serine but not between alanine and serine, indicating that serine was likely derived from glutamate via glutamate-phosphohydroxypyruvate transaminase.

However, using $^{13}$C-pyruvate and unlabeled glutamine as precursors we have found a remarkable formation of $^{13}$C-labeled alanine compared with serine (data not shown). Furthermore, our preliminary data using $^{13}$C-NMR indicate that OK cells utilized in the current studies do not produce glucose from pyruvate, however $^{31}$P-NMR analysis indicates accumulation of P-enol-pyruvate (data not shown). In addition, incubations with $^{13}$C-glucose and unlabeled glutamine as precursors resulted in higher formation of $^{13}$C-serine compared with $^{13}$C-alanine. These observations may occur as a result of incomplete metabolism of 3-P-glycerate, the glycolytic intermediate. Instead of being converted to pyruvate, 3-P-glycerate may be converted to 3-P-OH-pyruvate and, in the presence of glutamate, to serine. This possibility may occur due to lack or lower activity of phosphoglyceromutase in the OK cells. The observation that these OK cells do not form glucose from pyruvate support this hypothesis. However, further studies are required to determine this possibility.

In summary, the data thus far indicate that OK cells display glutaminase and GLDH activity as well as aminotransferase reactions. These pathways respond to alteration of media pH or supplementation of hormone to the incubation medium. It appears that these cells are not gluconeogenic, therefore, ammonia metabolism in various acid-base states can be studied independent of renal gluconeogenesis.

## References


1 Tannen RL, Sahai A: Biochemical pathways and modulators of renal ammoniagenesis. Miner Electrolyte Metab 1990;16:249–258.

2 Nissim I, States B, Yudkoff M, Segal S: Characterization of amino acid metabolism by cultured rat kidney cells: Study with $^{15}$N. Am J Physiol 1987;253:F1243–F1252.

3 Nissim I, States B: Ammoniagenesis by cultured human renal cortical epithelial cells: Study with $^{15}$N. Am J Physiol 1989;256:F187–F196.

4 Nissim I, Korzets Z: Ammonia metabolism in cultured renal cells. Miner Electrolyte Metab 1990;16:270–276.

5 Sahai A, Cole LA, Tannen RL: Pathways and regulation of ammoniagenesis by the LLC-$PK_1$ cells in culture. J Lab Clin Med 1989;114:285–293.

6 Sahai A, Laughrey E, Tannen RL: Relationship between intracellular pH and ammonia metabolism in LLC-$PK_1$ cells. Am J Physiol 1990;258:F103–F108.

7 Sahai A, Nissim I, Tannen RL: Pathways of acute pH regulation of ammoniagenesis in LLC-$PK_1$ cells: Study with [$^{15}$N]glutamine. Am J Physiol 1991;261:F481–F487.

8 Gstraunthaler G: Epithelial cells in tissue culture. Renal Physiol Biochem 1988;11:1–42.

9 Gstraunthaler G, Landauer F, Pfaller W: Ammoniagenesis in LLC-$PK_1$ cultures: Role of transamination. Am J Physiol 1992;263:C47–C54.

10 Fruelle T, Tong J, Gurthoys NP: Changes in rat renal glutaminase activity studies in vivo and in primary cultures of proximal convoluted tubular cells. Contrib Nephrol. Basel, Karger, 1985, vol 47, pp 150–156.

11 Handler JS, Perkins FM, Johnson JP: Studies of renal cell function using cell culture techniques. Am J Physiol 1980;238:F1–F9.
12 Handler JS: Studies of kidney cells in culture. Kidney Int 1986;36:208–215.
13 Bayliss JM, Socble JE, Thomasson KA, Trigg DJ, Buerkert JE: The effect of pH on ammonia secretion by proximal tubule monolayers in primary culture. Contrib Nephrol. Basel, Karger, 1988, vol 63, pp 21–27.
14 Margolis BL, Lifschitz MD: Ammonia production and amino acid metabolism by rat renal papillary epithelial cells in culture. J Biol Chem 1985;260:501–507.
15 Pollock AS, Warnock DG, Streuler GJ: Parathyroid hormone inhibition of $Na^+$-$H^+$ antiporter activity in cultured renal cell lines. Am J Physiol 1986;250:F217–F225.
16 Montrose MH, Murer H: Regulation of intracellular pH by cultured opossum kidney cells. Am J Physiol 1990;259:C110-C120.
17 Abraham MI, Mcateer JA, Kempson SA: Insulin stimulates phosphate transport in opossum kidney epithelial cells. Am J Physiol 1990;258:F1592–F1598.
18 Murphy TJ, Bylund DB: Characterization of alpha-2-adrenergic receptors in the OK cells, and opossum kidney cell lines. J Pharmacol Exp Ther 1988;244:571–578.
19 Guntupalli J, Rogers A, Bourke E: Effect of insulin on renal phosphorus handling in the rat: Interaction with PTH and nicotinamide. Am J Physiol 1985;249:F610–F618.
20 Nissim I, Wehrli S, States B, Nissim I, Yudkoff M: Analysis and physiological implications of renal 2-oxoglutaramate metabolism. Biochem J 1991;277:33–38.

Itzhak Nissim, PhD, Division of Biochemical Development and Molecular Diseases,
Children's Hospital of Philadelphia, 34th Street & Civic Center Boulevard,
Philadelphia, PA 19104 (USA)

Tizianello A, Baverel G, Endou H, Schoolwerth AC, O'Donovan DJ (eds):
Renal Ammoniagenesis and Interorgan Cooperation in Acid-Base Homeostasis.
Contrib Nephrol. Basel, Karger, 1994, vol 110, pp 41–45

# Effect of Valproate and of Two of Its Metabolites on Renal Substrate Uptake and Ammoniagenesis in the Rat in vivo

*Mohamed Elhamri, Bernard Ferrier, Mireille Martin, Gabriel Baverel*

Laboratoire de Physiopathologie Métabolique et Rénale et de Spectroscopie RMN (CNRS EP 18), Faculté de Médecine Alexis-Carrel, Lyon, France

Administration of sodium valproate (VPA), a widely used and very effective antiepileptic drug [1], often induces a moderate hyperammonemia with extremely rare clinical signs of liver dysfunction. Experiments performed in vivo both in man and rat point to the kidneys as organs responsible, at least in part, for the valproate-induced hyperammonemia; these experiments demonstrated an increase in the renal uptake and metabolism of glutamine resulting in a stimulation of the renal venous release of ammonia [2–5].

The aim of the present study was to test in the rat the hypothesis that VPA does so by inhibiting the renal uptake and metabolism of substrates such as lactate, fatty acids, ketone bodies and α-ketoglutarate, which are known inhibitors of renal ammoniagenesis [6–9]. For this, we have used an in vivo rat kidney preparation and studied the effect of VPA and of two of its metabolites (4-en-VPA and 2-en-VPA) on the renal production of ammonia and on the renal handling of glutamine, glutamate, glucose, lactate, free fatty acids, ketone bodies, citrate, glycerol and α-ketoglutarate.

## Materials and Methods

Male rats fasted overnight, obtained from Iffa-Credo (St. Germain sur l'Arbresle, France), weighing 200–250 g, were used. The surgical preparation was as reported in a previous publication [10]. Blood flow, urine flow, glomerular filtration rate as well as consumption (or production) and excretion of the 11 substances studied have been determined precisely for the left kidney both during a 25-min control period and a 25-min experimental period starting 15 min after the intravenous administration of a bolus of 200 mg/kg b.w. of VPA or of one of its metabolites thanks to methods previously described [11].

*Table 1.* Effect of VPA, 4-en-VPA and 2-en-VPA on the overall function of the left kidney of 24-hour fasted rats

| Animals | GFR $ml \cdot kg^{-1} \cdot min^{-1}$ | Urine flow $ml \cdot kg^{-1} \cdot min^{-1}$ | PAH extraction, % | RBF $ml \cdot kg^{-1} \cdot min^{-1}$ |
|---|---|---|---|---|
| Control | 3.8 ± 0.2 | 0.22 ± 0.03 | 79.6 ± 1.1 | 26.7 ± 1.9 |
| VPA (n = 7) | 3.6 ± 0.2 | 0.26 ± 0.03 | 73.5 ± 2.4* | 21.7 ± 0.6 |
| Control | 4.2 ± 0.4 | 0.39 ± 0.04 | 77.0 ± 1.9 | 29.3 ± 3.2 |
| 4-en-VPA (n = 4) | 4.1 ± 0.6 | 0.35 ± 0.04 | 72.2 ± 1.6* | 24.5 ± 3.0 |
| Control | 4.1 ± 0.2 | 0.23 ± 0.04 | 79.6 ± 2.1 | 28.9 ± 2.0 |
| 2-en-VPA (n = 5) | 4.5 ± 0.4 | 0.40 ± 0.04* | 69.8 ± 3.2* | 26.2 ± 1.1 |

The data are means ± SEM. n = Number of rats of each group. Statistical difference was measured by the paired Student's t test against the appropriate control (* $p < 0.05$).

GFR = Glomerular filtration rate; PAH = *p*-aminohippurate; RBF = renal blood flow.

## Results

Table 1 shows that neither VPA nor any of its two metabolites altered the GFR and renal hemodynamics. The urine flow was increased only by 2-en-VPA. By contrast, the three compounds significantly reduced renal PAH extraction.

It can be seen in table 2 that administration of both VPA and 4-en-VPA, but not of 2-en-VPA, caused a hyperammonemia. The three compounds increased the blood glutamate concentration; this effect was accompanied by an increased renal glutamate uptake. Although the blood concentration of glutamine remained unchanged, the renal uptake of this amino acid was stimulated by VPA and 4-en-VPA and, to a smaller extent, by 2-en-VPA.

Increased uptake of glutamate and glutamine were associated with a marked stimulation of the renal production and excretion of ammonia after the administration of VPA and 4-en-VPA. The stimulation of the renal production of ammonia was much less pronounced in the presence of 2-en-VPA and, in contrast with what was observed with VPA and 4-en-VPA, did not result in an increased renal venous release of ammonia (data not shown).

The three compounds induced a fall of the blood concentration and renal uptake of β-hydroxybutyrate, but the blood concentration and renal uptake of acetoacetate was diminished only by the administration of VPA and 4-en-VPA. Although only 2-en-VPA caused a diminution of the plasma FFA concentration, the renal FFA uptake was considerably reduced by VPA and even

*Table 2.* Effect of VPA, 4-en-VPA and 2-en-VPA on substrate handling by the left kidney of 24-hour fasted rats

| Substrate | VPA | | 4-en-VPA | | 2-en-VPA | |
|---|---|---|---|---|---|---|
| | arterial blood concentration m*M* | uptake $\mu mol \cdot kg^{-1} \cdot min^{-1}$ | arterial blood concentration m*M* | uptake $\mu mol \cdot kg^{-1} \cdot min^{-1}$ | arterial blood concentration m*M* | uptake $\mu mol \cdot kg^{-1} \cdot min^{-1}$ |
| *Glutamine* | | | | | | |
| Control | 0.436±0.023 | 0.75±0.24 | 0.490±0.032 | 0.75±0.29 | 0.458±0.037 | 0.55±0.36 |
| Experimental | 0.469±0.015 | 2.29±0.19* | 0.523±0.062 | 2.62±0.41* | 0.493±0.039 | 1.50±0.11* |
| *Glutamate* | | | | | | |
| Control | 0.168±0.014 | 0.64±0.13 | 0.200±0.016 | –0.08±0.15 | 0.162±0.024 | –0.21±0.10 |
| Experimental | 0.242±0.026* | 1.31±0.23* | 0.293±0.052* | 1.16±0.44* | 0.238±0.028* | 0.59±0.19* |
| *Ammonia* | | | | | | |
| Control | 0.078±0.009 | –3.48±0.33 | 0.078±0.019 | –3.44±0.47 | 0.050±0.010 | –3.23±0.54 |
| Experimental | 0.096±0.007* | –5.20±0.52* | 0.104±0.024* | –5.28±0.27* | 0.060±0.009 | –4.12±0.52* |
| *β-Hydroxybutyrate* | | | | | | |
| Control | 1.038±0.165 | 2.69±0.58 | 1.237±0.172 | 2.17±0.61 | 1.115±0.130 | 2.60±0.62 |
| Experimental | 0.205±0.051* | –0.03±0.2* | 0.285±0.018* | –0.22±0.29* | 0.590±0.07* | 0.54±0.68* |
| *Acetoacetate* | | | | | | |
| Control | 0.965±0.093 | 2.29±0.77 | 0.887±0.050 | 2.80±0.54 | 0.966±0.117 | 2.94±0.41 |
| Experimental | 0.608±0.070* | 0.70±0.44* | 0.572±0.046* | –0.13±0.60* | 0.815±0.085 | 2.72±0.79 |
| *Plasma FFA* | | | | | | |
| Control | 0.808±0.074 | 1.32±0.33 | 0.686±0.029 | 0.94±0.10 | 0.880±0.016 | 1.35±0.69 |
| Experimental | 0.923±0.124 | 0.51±0.30* | 0.879±0.257 | –0.09±0.05* | 0.454±0.026* | –0.19±0.03* |
| *Lactate* | | | | | | |
| Control | 0.434±0.022 | 2.11±0.81 | 0.590±0.090 | 2.71±0.68 | 0.598±0.055 | 2.18±0.40 |
| Experimental | 1.005±0.185* | 1.86±0.42 | 0.981±0.113* | 2.45±1.27 | 0.852±0.039* | 2.47±0.92 |
| *Glycerol* | | | | | | |
| Control | 0.105±0.009 | 1.58±0.24 | 0.108±0.006 | 1.40±0.19 | 0.106±0.008 | 1.27±0.27 |
| Experimental | 0.120±0.010 | 1.61±0.24 | 0.143±0.015* | 1.85±0.27* | 0.107±0.018 | 1.17±0.16 |
| *Citrate* | | | | | | |
| Control | 0.168±0.009 | 1.59±0.21 | 0.146±0.019 | 1.19±0.17 | 0.144±0.013 | 1.49±0.16 |
| Experimental | 0.153±0.010* | 1.21±0.21 | 0.128±0.019 | 1.13±0.23 | 0.121±0.015* | 1.20±0.10 |
| *α-Ketoglutarate* | | | | | | |
| Control | 0.020±0.002 | 0.18±0.03 | 0.022±0.002 | 0.09±0.03 | 0.022±0.002 | 0.17±0.03 |
| Experimental | 0.017±0.002 | –0.02±0.02* | 0.023±0.004 | –0.05±0.05* | 0.015±0.002 | 0.08±0.05* |

The data obtained are means±SEM for seven experiments. Results for metabolite uptake or production (a negative uptake) were calculated as described under Methods. Statistical difference was measured by the paired Student's t test against the control without valproate. * $p < 0.05$.

suppressed by 4-en-VPA and 2-en-VPA. Despite an elevation of the blood lactate concentration, the renal lactate uptake remained unchanged in the presence of the three compounds.

Table 2 also shows that the renal uptake of α-ketoglutarate was inhibited in the presence of 2-en-VPA, and converted into a net production in the presence of VPA and 4-en-VPA. The only other significant change observed was an increased glycerol uptake caused by 4-en-VPA. The blood concentration and renal handling of glucose were not altered after the administration of VPA or its two metabolites.

## Discussion

This study establishes that, in the rat, the hyperammonemia known to be caused by the administration of VPA [2, 4, 5], is also observed after the administration of 4-en-VPA but not 2-en-VPA. It also supports the view expressed by Warter et al. [2, 3] and later by other authors [4, 5] that this hyperammonemia results, at least in part, from an increased renal venous release of ammonia.

Our results demonstrating an increased renal uptake of glutamine in the presence of VPA are in agreement with previous findings of other authors [2, 3, 5]; that this is also the case after the administration of 4-en-VPA and 2-en-VPA represents a new piece of information. In addition, our study suggests that the stimulation of the renal production of ammonia probably results from an increased metabolism not only of glutamine but also of glutamate.

It should be underlined that this study demonstrates that administration of VPA and of the two VPA metabolites studied inhibits to various extents the renal uptake of ketone bodies, fatty acids and α-ketoglutarate. Because the latter circulating substrates are known to inhibit renal ammoniagenesis [6–9], our results strongly suggest that their reduced renal uptake caused by VPA, 4-en-VPA and 2-en-VPA was responsible, at least in part, for the stimulation of the renal production of ammonia observed after the administration of these pharmacological agents.

Finally, calculations from the data of table 2 reveal that VPA, 4-en-VPA and 2-en-VPA reduced the total amount of substrate carbons removed and presumably metabolized by the kidney to approximately the same extents. Despite this, the stimulation of renal ammoniagenesis was not of the same magnitude in the presence of these three compounds. Since VPA and 4-en-VPA were more potent stimulators of renal ammoniagenesis than 2-en-VPA, it is likely that, in addition to reducing the renal uptake and metabolism of inhibitors of renal ammoniagenesis, they stimulated the latter process by some other unidentified mechanism.

## References


1 Simon D, Penry JK: Sodium di-*n*-propylacetate in the treatment of epilepsy: A review. Epilepsia 1975;16:549–573.

2 Warter JM, Imler M, Marescaux C, Chabrier G, Rumbach L, Micheletti G, Krieger J: Sodium valproate-induced hyperammonemia in the rat: Role of the kidney. Eur J Pharmacol 1983;87: 177–182.

3 Warter JM, Marescaux C, Chabrier G, Rumbach L, Micheletti G, Imler M: Métabolisme rénal de la glutamine chez l'homme au cours du traitement par le valproate de sodium. Rev Neurol (Paris) 1984;140:370–371.

4 Ferrier B, Martin M, Baverel G: Valproate-induced stimulation of renal ammonia production and excretion in the rat. J Clin Chem Clin Biochem 1988;26:65–67.

5 Culebras M, Doval M, Rengel M, Gougoux A, Vinay P, Lopez-Novoa JM: Effect of valproate infusion in vivo on renal metabolism in fasted rats. Biochem Pharmacol 1989;38:3631–3634.

6 Lemieux G, Vinay P, Baverel G, Brière R, Gougoux A: Relationship between lactate and glutamine metabolism in vitro by the kidney: Differences between dog and rat and importance of alanine synthesis in the dog. Kidney Int 1979;16:451–458.

7 Baverel G, Michoudet C, Martin G: Role of fatty acids in simultaneous regulation of flux through glutaminase and glutamine synthetase in rat kidney cortex; in Häussinger D, Sies H (eds): Glutamine Metabolism in Mammalian Tissues. Berlin, Springer, 1984, pp 187–202.

8 Goldstein L, Solomon RJ, Perlman DF, McLaughlin PM, Taylor MA: Ketone body effects on glutamine metabolism in isolated kidneys and mitochondria. Am J Physiol 1982;243:F181–F187.

9 Churchill PC, Malvin RL: Relation of renal gluconeogenesis to ammonia production in the rat. Am J Physiol 1970;218:353–357.

10 Ferrier B, Martin M, Baverel G: Reabsorption and secretion of α-ketoglutarate along the rat nephron: A micropuncture study. Am J Physiol 1985;248:F404–F412.

11 Elhamri M, Martin M, Ferrier B, Baverel G: Substrate uptake and utilization by the kidney of fed and starved rats in vivo. Renal Physiol Biochem 1993;16:311–324.



Dr. G. Baverel, Université Claude Bernard, U.E.R., Faculté de Médecine Alexis-Carrel, Rue Guillaume Paradin, F–69372 Lyon Cedex 2 (France)

Tizianello A, Baverel G, Endou H, Schoolwerth AC, O'Donovan DJ (eds):
Renal Ammoniagenesis and Interorgan Cooperation in Acid-Base Homeostasis.
Contrib Nephrol. Basel, Karger, 1994, vol 110, pp 46–53

# Bicarbonate-Dependent Betaine Synthesis in Rat Kidney[1]

*Gilbert Moeckel, Hans Günther Dasser, Tian Jun Chen, Michael Schmolke, Walter Georg Guder*

Institute of Clinical Chemistry, Bogenhausen Hospital, Munich, FRG

Trimethylamine betaine has been described as one of the major organic osmolytes which accumulate along the corticopapillary axis in mammalian kidney [1, 2]. Betaine is known to be formed from choline in liver and kidney [3, 4] and metabolized to glycine by consecutive transmethylation reactions [5]. In a previous study in rabbit kidney we have shown that betaine synthesis is localized mainly in the proximal tubule [4]. In addition, Grossmann and Hebert [6] documented choline dehydrogenase activity in the inner medulla of rabbit kidney. The latter finding was confirmed by Schmolke et al. [2] in rat kidney inner medullary tubules.

Betaine concentrations exhibit a different distribution pattern from that of choline dehydrogenase activity with concentration maxima in the outer medulla and the papillary tip, respectively [7].

From the apparent discrepancies in the intrarenal distribution pattern of betaine synthesis and content the question regarding the relative contribution of transport versus biosynthesis to betaine accumulation was raised [1]. A sodium-dependent betaine transporter has been described in renal medullary cells [8] and recently cloned in Xenopus oocytes by Yamauchi et al. [9].

In the present study we compared the rates of betaine synthesis with those of betaine uptake in tubules from defined rat kidney sections. In addition, the intracellular localization and reaction mechanism of renal betaine synthesis was investigated.

[1] This work has been supported by the Deutsche Forschungsgemeinschaft (Gu 82/3-3).

## Materials and Methods

*Homogenates and Tubules*

Rat kidneys were excised after decapitation and sliced freehand with a scalpel into papillary tip, papillary base, inner medulla, outer medulla (outer stripe and inner stripe) and outer cortex and inner cortex. 10% (w/v) homogenates were prepared in Krebs-Henseleit-bicarbonate buffer, pH = 7.4, with a Teflon-glass homogenizer [10]. Mitochondrial fractions were prepared from renal cortex by differential centrifugation as described elsewhere [6, 11]. Isolated tubule suspensions were prepared from kidney cortex [12], outer [13] and inner medulla [10] as described previously.

*Betaine Uptake*

Tubule suspensions were incubated for 60 min at 37 °C with 0.1 mmol/l $^{14}C$-betaine. After separation of the tubules, radioactivity was determined in the pellet fraction by liquid scintillation counting and by HPLC analysis [14].

*Choline Metabolism*

In order to quantitate the rate of betaine formation from choline, homogenates, tubule suspensions and mitochondrial fractions were incubated in KRB buffer with 1 mmol/l unlabeled and/or methyl-$^{14}C$-choline [4, 15]. Choline was determined enzymatically as described in detail by Wirthensohn and Guder [4]. Betaine and choline phosphate were identified after separation by HPLC [14] using a through-flow HPLC radioactivity detector (LB 506 B; Fa. Berthold, Wildbad, FRG) and a liquid scintillation counter (LS 1801; Beckman Instruments, Inc., Irvine, Calif., USA).

## Results

*Bicarbonate Dependence of Choline Oxidation*

Suspensions of proximal tubules incubated in Krebs-Henseleit buffer metabolized choline at a rate of $218 \pm 30$ (n = 7) mmol/kg protein/h similar to the extent observed in homogenates [15]. When the incubation was performed in sucrose or Tris buffers, choline uptake was negligible. Omitting the individual components of the Krebs-Henseleit buffer, bicarbonate turned out to be the essential component for choline uptake and betaine formation. Bicarbonate concentrations up to 20 mmol/l increased choline uptake with an apparent half maximal stimulatory concentration of 8 mmol/l and no further stimulation above 20 mmol/l. Betaine formation which accounted for 77% of choline uptake in Krebs-Henseleit buffer was likewise dependent on the presence of millimolar amounts of bicarbonate (fig. 1). This bicarbonate dependence of betaine synthesis was also observed in preparations of isolated mitochondria or inner mitochondrial membranes [15]. More than 80% of cortical bicarbonate-dependent betaine formation resided with the inner mitochondrial membrane upon tissue fractionation. A similar bicarbonate dependence was found in inner medullary homogenates.

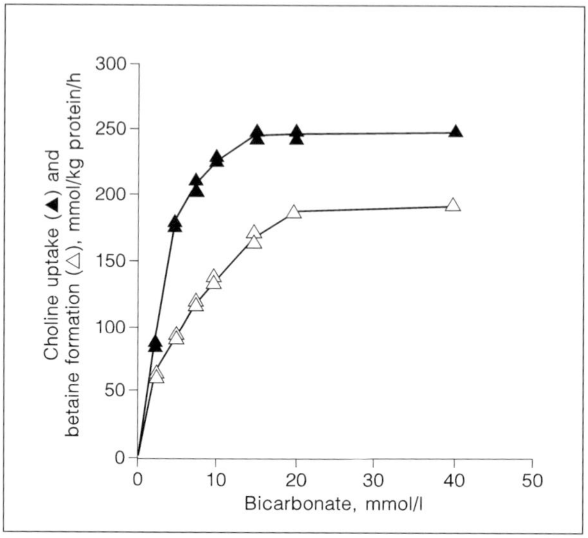


*Fig. 1.* Bicarbonate dependence of choline metabolism in rat kidney cortex homogenate. 10% (w/v) homogenates from kidney cortex were incubated in Krebs-Henseleit-bicarbonate buffer [15], with increasing concentrations of sodium bicarbonate, ranging from 2.5 to 40 mmol/l in the incubation medium. Choline was determined enzymatically [4] and betaine formation by HPLC [2].

*Effect of NAD and SH Donors on Betaine Formation*

From the previous literature, choline dehydrogenase is thought to form betaine aldehyde in a NAD-dependent reaction [3, 6]. Betaine aldehyde on the other hand is further metabolized to betaine by an unspecific aldehyde dehydrogenase which also uses NAD [11]. In accordance with this concept, 5 mmol/l NAD in the incubation medium of kidney cortex or inner medullary homogenates was found to stimulate betaine formation severalfold (fig. 2). Interestingly this stimulatory effect was even higher in the additional presence of 5 mmol/l glutathione, which by itself had only 50% of the stimulatory effect of NAD alone. A similar stimulatory effect was obtained when dithioerythritol (DTE) was added instead of GSH (not shown). NADP caused an increase of betaine formation comparable to that of NAD (fig. 3). In contrast to NAD however, glutathione was not stimulatory in the presence of NADP. From this experience, NAD and GSH were included in all further experiments to assure maximum betaine formation.

Earlier studies used oxygen uptake as a measure of choline oxidation [16], indicating a link of this enzyme to mitochondrial respiration. In accordance

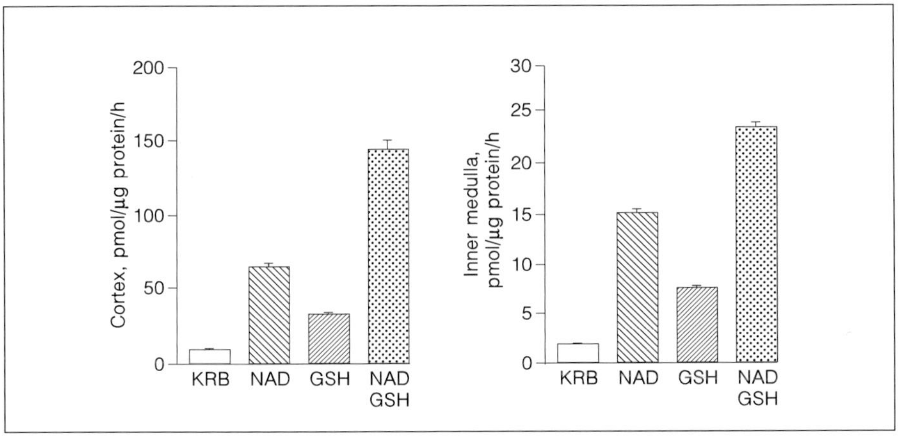


*Fig. 2.* Cofactor dependence of betaine formation in rat kidney. Cortex and inner medulla homogenates (1 g/l) were prepared as described in the Methods and incubated for 1 h with [14]C-methylcholine as described before [4]. NAD and GSH were added to the incubation at a concentration of 5 mmol/l. Betaine formation was determined after HPLC by liquid scintillation [2].

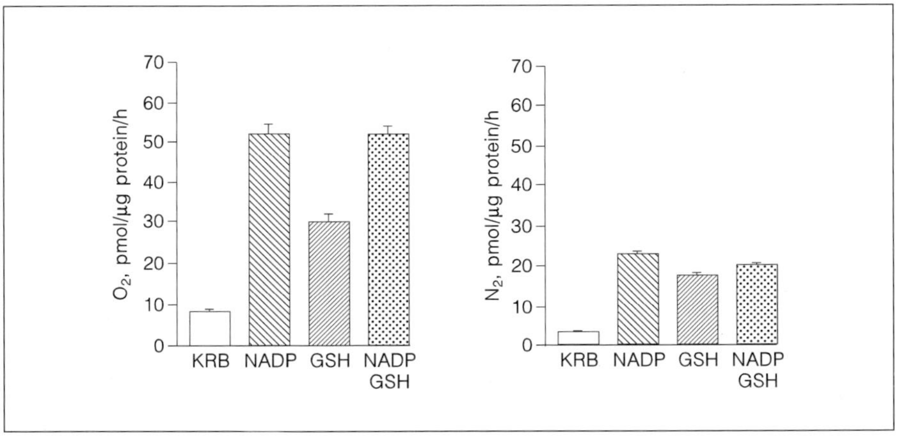


*Fig. 3.* Oxygen dependence of betaine formation in rat kidney cortex homogenates. Betaine synthesis from [14]C-choline was determined under oxygen and nitrogen atmosphere. Cortex homogenate preparation and incubation was performed as described in the legend to figure 2, except that NADP was used as cofactor. Samples were gassed for several minutes either with carbogen (95% $O_2$/5% $CO_2$) [4], or a mixture of 90% $N_2$/10% $CO_2$ for the same period of time. [14]C-betaine was detected after ion exchange chromatography [4] by liquid scintillation.

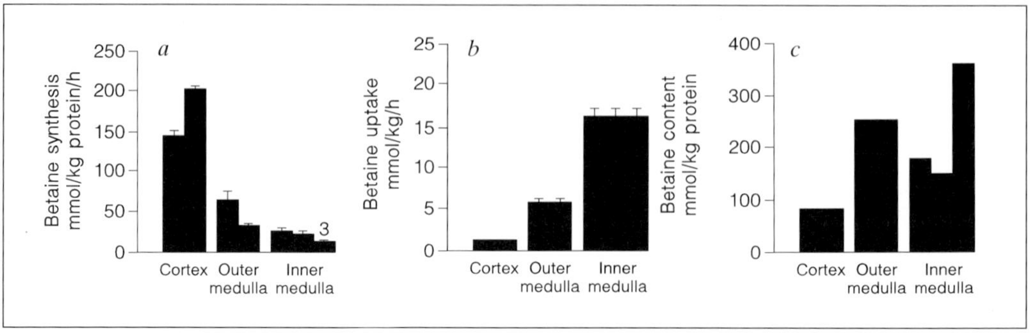


*Fig. 4.* Distribution of betaine synthesis, betaine uptake and concentration along the corticopapillary axis of rat kidney. *a* Betaine synthesis from $^{14}C$-methylcholine was determined in homogenates of seven different rat kidney regions. *b* Uptake of $^{14}C$-betaine in tubule suspensions from cortex, outer and inner medulla. *c* Betaine concentration along the corticopapillary axis, measured by HPLC in rat kidney in vivo [from 2].

with this assumption, betaine formation rate was decreased by 60% in the absence of oxygen either with or without NADP or GSH (fig. 3). All following experiments were therefore performed under high oxygen atmosphere (5% $CO_2$/95% $O_2$).

*Distribution of Betaine Synthesis and Uptake along the Corticopapillary Axis*

When betaine synthesis was quantitated along the corticopapillary axis, the results shown in figure 4a were obtained. Betaine synthesis was highest in the cortex and decreased towards the papillary tip with rates ranging from 199 ± 11 mmol/kg protein/h in the cortical inner zone to 15.8 ± 1.2 mmol/kg protein/h in IM 3 segment of the inner medulla.

In contrast, tubular betaine uptake of different zones of the kidney exhibited a completely different distribution pattern. Active betaine accumulation was nearly absent in cortical tubules, amounted to 7 mmol/kg protein/h in the outer medulla and increased further in the inner medulla to 16.2 mmol/kg protein/h (fig. 4b).

For comparison, figure 4c gives the respective betaine concentrations in different kidney zones found under in vivo conditions [17].

## Discussion

*Possible Mechanisms of Betaine Formation*

Betaine is found as a product of choline metabolism in bacteria, plants and throughout the animal kingdom [for review, see 17]. A choline oxidase using oxygen and forming $H_2O_2$ has been isolated from several species [18]. In the first reaction step this enzyme forms betaine aldehyde, which is further oxidized to betaine, resulting in 2 mol $H_2O_2$/mol choline oxidized. Zhang et al. [16] were unsuccessful to find any $H_2O_2$ formation in liver mitochondria oxidizing choline, but confirmed the earlier finding of oxygen uptake coupled to choline oxidation. In mammalian liver [3] and kidney [6] a NAD-dependent choline dehydrogenase has been described. In addition, an unspecific betaine aldehyde dehydrogenase, that also seems to be NAD-dependent, was characterized [3]. Oxidation of 1 mol choline could therefore result in 2 mol NADH. Our observations indicate that the mechanism of choline oxidation is dependent on oxygen and NAD (NADP). This can be interpreted in a way that NAD is used as a cofactor in at least one of the subsequent reactions (choline dehydrogenase or aldehyde dehydrogenase). Oxygen may either act as a substrate or as an oxidant channelling electrons to the respiratory chain. Our observations that the choline uptake of inner mitochondrial membranes is abolished in the presence of cyanide [15; data not shown here] point to a tight coupling of choline oxidation to the respiratory chain in this cell subfraction. A similar conclusion has been drawn by Zhang et al. [16] from experiments with liver mitochondria.

In addition, a stimulatory effect of reduced glutathione was observed. The positive effect of glutathione and dithiothreitol may be explained by the protective function of these compounds on essential SH groups for enzymatic activity. Alternatively, glutathione may lead to inhibition of cathepsins [19] which are known to be highly active in the kidney tubule [20] and may destroy choline dehydrogenase in vitro. Moreover, glutathione is the substrate of glutathione peroxidase and may inactivate $H_2O_2$ formed in homogenates. Due to this mechanism, glutathione may prevent peroxidative destruction of the enzyme. Purification of the enzyme and further studies on the nature of the stimulatory effects observed are needed before final conclusions can be drawn. In conclusion, the exact mechanism of choline oxidation in rat kidney mitochondria seems less clearly defined than described in previous papers.

This also holds for the observation that betaine formation was dependent on the presence of bicarbonate. This is the first time that this effect is described in the literature. Interestingly, the effect was maintained in isolated mitochondrial membranes, thus excluding indirect effects of this anion by bicarbonate-dependent cellular or mitochondrial transporters. Surprisingly the greatest effect is observed in the acidotic range of bicarbonate. Further studies of betaine formation depending on the acid-base state of the organism may elucidate the possible physiological implications of this observation.

*Contribution of Betaine Synthesis versus Uptake to Accumulation of This Osmolyte in Various Kidney Zones*

Highest rates of betaine synthesis were found in the proximal tubule in accordance with earlier findings in rabbit kidney [4]. This contrasts with the intrarenal distribution pattern of betaine under antidiuretic conditions, where highest concentrations are found in the papilla and the outer medulla (fig. 4c). The present results indicate that synthesis and transport may both contribute to the accumulation of betaine in the kidney. Proximal tubules when incubated in isotonic saline, released most of their newly formed betaine into the extracellular space [15]. This explains why in the presence of extracellular betaine cortical tubules do not accumulate this osmolyte to any extent.

Outer and inner medullary tubules on the other hand accumulate betaine by an osmolality-dependent mechanism leading to highest accumulation in the inner medulla (fig. 4b). Since betaine is not degraded to any significant extent in this region of the kidney [Dasser et al., unpubl. observation], transport may be the major mechanism of betaine accumulation in the renal papilla. By comparing the rates of synthesis and transport at physiological plasma choline and betaine concentrations, one may calculate the relative contributions of both mechanisms to tubular betaine accumulation. On the other hand, it is not known whether proximal tubular betaine can reach the papilla via the tubular lumen. The finding that newborns excrete high amounts of betaine into the urinary space [5] support the assumption that betaine filtered by the glomerulus or secreted by the proximal tubule may be taken up by distal nephron cells according to the interstitial tonicity directed on them. In addition to these mechanisms, an osmolality-dependent betaine release has to be considered in regulating tubular betaine concentration. In acute water diuresis [14] and furosemide treatment [21] and in isolated inner medullary tubules, a release of betaine into the extracellular space has been observed.

## References

1 Beck FX, Schmolke M, Guder WG: Osmolytes. Curr Opin Nephrol Hypertens 1992;1:43–52.
2 Schmolke M, Bornemann A, Guder WG: Distribution and regulation of organic osmolytes along the nephron; in Koide H, Endou H, Kurokawa H (eds): Cellular and Molecular Biology of the Kidney. Contrib Nephrol. Basel, Karger, 1991, vol 95, pp 255–263.
3 Wilken DR, McMacken ML: Choline and betaine aldehyde oxidation by rat liver mitochondria. Biochim Biophys Acta 1970;216:305–317.

4 Wirthensohn G, Guder WG: Studies on renal choline metabolism and phosphatidylcholine synthesis; in Morel F (ed): Biochemistry of Kidney Function. Amsterdam, Elsevier, 1982, pp 119–128.
5 Davies SEC, Chalmers RA, Woolf DA, Iles RA: Betaine metabolism in the human neonate. Biochem Soc Trans 1988;16:788–789.
6 Grossmann EB, Hebert SC: Renal inner medullary choline dehydrogenase activity: Characterization and modulation. Am J Physiol 1989;256:F107–F112.
7 Guder WG, Beck FX, Schmolke M: Regulation and localization of organic osmolytes in mammalian kidney. Klin Wochenschr 1990;68:1091–1095.
8 Nakanishi T, Turner RJ, Burg MB: Osmoregulation of betaine transport in mammalian renal medullary cells. Am J Physiol 1990;258:F1061–F1067.
9 Yamauchi A, Uchida S, Kwon HM, Preston AS, Robey RB, Garcia-Perez A, Burg MB, Handler JS: Cloning of a $Na^+$- and Cl-dependent betaine transporter that is regulated by hypertonicity. J Biol Chem 1992;267:649–652.
10 Wirthensohn G, Lefrank S, Schmolke M, Guder WG: Regulation of organic osmolyte concentration in tubules from rat renal inner medulla. Am J Physiol 1989;256:F128–F135.
11 Haubrich DR, Gerber NH: Choline dehydrogenase: Assay, properties and inhibitors. Biochem Pharmacol 1981;30:2993–3000.
12 Guder WG: Stimulation of renal gluconeogenesis by angiotensin II. Biochim Biophys Acta 1979;584:507–519.
13 Schmolke M, Guder WG: Metabolic regulation of organic osmolytes in tubules from rat renal inner and outer medulla. Renal Physiol Biochem 1989;12:347–358.
14 Beck FX, Schmolke M, Guder WG, Dörge A, Thurau K: Osmolytes in renal medulla during rapid changes in papillary tonicity. Am J Physiol 1992;262:F849–F856.
15 Dasser HG: Untersuchungen zum Cholinstoffwechsel der Niere; Thesis, Ludwig-Maximilians-Universität, Munich 1993.
16 Zhang J, Blusztajn JK, Zeisel SH: Measurement of the formation of betaine aldehyde and betaine in rat liver mitochondria by a high pressure liquid chromatography-radioenzymatic assay. Biochim Biophys Acta 1992;1117:333–339.
17 Yancey PH, Clark ME, Hand SC, Bowlus RD, Somero GN: Living with water stress: The evolution of osmolyte systems. Science 1982;217:1214–1222.
18 Ikuta S, Imamura S, Misaki H, Horiuti Y: Purification and characterization of choline oxidase from *Arthrobacter globiformis*. J Biochem 1977;82:1741–1749.
19 Katunuma N: Roles of glutamine and glutathione in terminal amino-nitrogen metabolism; in Endou H, Schoolwerth AC, Baverel G, Tizianello A (eds): Molecular Aspects of Ammoniagenesis. Contrib Nephrol. Basel, Karger, 1991, vol 92, pp 1–12.
20 Olbricht CJ: Distribution of cathepsin B and L in the kidney and their role in tubular protein reabsorption. Eur J Clin Chem Clin Biochem 1992;30:675–681.
21 Nakanishi T, Yamada T, Iida T, Uyama O, Sugita M: Betaine excretion increases by water loading after dehydration; in Koide H, Endou H, Kurokawa K (eds): Cellular and Molecular Biology of the Kidney. Contrib Nephrol. Basel, Karger, 1991, vol 95, pp 264–271.

Prof. Dr. Walter G. Guder, Institut für Klinische Chemie, Städtisches Krankenhaus München-Bogenhausen, Englschalkinger Strasse 77, D–81925 München (FRG)

Tizianello A, Baverel G, Endou H, Schoolwerth AC, O'Donovan DJ (eds):
Renal Ammoniagenesis and Interorgan Cooperation in Acid-Base Homeostasis.
Contrib Nephrol. Basel, Karger, 1994, vol 110, pp 54–59

# $NH_4^+$ and $NH_3$ Permeabilities of Henle's Loop Segments

*M.A. Knepper, M.F. Flessner, R. Mejia, C.-L. Chou*

Laboratory of Kidney and Electrolyte Metabolism, National Heart, Lung, and Blood Institute, National Institutes of Health, Bethesda, Md., USA

Regulated excretion of ammonium is crucial to the maintenance of acid-base balance. Ammonium excretion must be maintained at a level that is appropriate for the maintenance of acid-base balance regardless of the rate of water excretion. Hence, ammonium (like other solutes) must be concentrated in the urine when water excretion is limited [1]. Ammonium is concentrated in the urine as a result of two 'trapping' processes: countercurrent trapping in the renal medulla (which involves mainly $NH_4^+$ transport) and diffusion trapping in the collecting ducts (which involves passive $NH_3$ diffusion) [2–5]. During water-conserving states, these two processes act in unison to increase the ammonium concentration in the final urine to levels far above those achievable based only on pH gradient-mediated diffusion trapping.

Countercurrent trapping occurs by countercurrent multiplication of ammonium between the two limbs of Henle's loops and is responsible for raising the medullary interstitial ammonium concentration above that of the renal cortex [2–5]. The countercurrent multiplication process is driven by active $NH_4^+$ transport from the medullary thick ascending limb of Henle's loop [6–8]. While the countercurrent multiplication mechanism that generates a high ammonium concentration in the outer medulla is well understood, the details of the concentrating process in the inner medulla remain to be worked out. This is in part owing to the level of difficulty involved in studying the thin limb segments of Henle's loop.

The purpose of this short review is to describe recent work in our laboratory to determine the passive permeability properties of loop of Henle segments, and to consider the significance of these data to the process that concentrates ammonium in the urine.

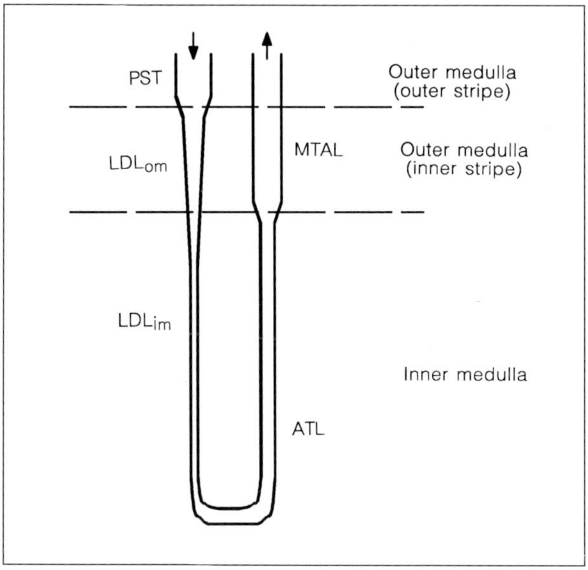


*Fig. 1.* Terminology for renal tubule segments included in long Henle's loop. PST, proximal straight tubule; $LDL_{om}$, long loop descending limb in outer medulla; $LDL_{im}$, long loop descending limb in inner medulla; ATL, ascending thin limb; MTAL, medullary thick ascending limb. The PST, which extends through the outer stripe of the outer medulla and medullary ray, has also been termed the 'thick descending limb'. The $LDL_{im}$ is itself heterogeneous; the $LDL_{im}$ results presented here are in tubules from the middle region of the inner medulla. Ultrastructural studies of perfused chinchilla thin limb segments have demonstrated that the $LDL_{om}$ has a type II thin limb epithelium, and that the $LDL_{im}$ segments used in this study have a type III thin limb epithelium [Chou et al., unpubl. observations].

## Methods

The methods utilized for these studies have been described in detail previously [7, 9–12] and only a brief general description will be given here. The measurements of $NH_3$ and $NH_4^+$ permeability are carried out in isolated perfused tubules using continuous flow fluorometry to measure total ammonia concentrations [13]. The permeabilities are measured by two general methods. In one method, a gradient for either $NH_3$ or $NH_4^+$ is set across the epithelium by appropriate adjustments of total ammonia concentration and pH. The permeabilities are calculated from the measured total ammonia fluxes and the calculated $NH_3$ and $NH_4^+$ gradients. In the other method, gradients for both $NH_3$ and $NH_4^+$ are set up across the epithelium and the measured flux of total ammonia is partitioned into $NH_3$ and $NH_4^+$ fluxes using balance equations for both total ammonia and net acid.

For the studies described here, renal tubule segments have been microdissected from rats, chinchillas, and rabbits. The thin limb segments of Henle's long loop were identified using criteria developed by Chou and Knepper [14]. The terminology used for the renal tubule segments is shown in figure 1. There are two general classes of nephrons in most mammalian species: those with short loops of Henle (which bend in the outer medulla), and those with long loops of Henle (which bend in the inner medulla). This report is concerned with the permeability properties of the long loops.

## Results and Discussion

Although the concept of countercurrent multiplication of ammonium is well established based on the findings that $NH_4^+$ is actively reabsorbed from the medullary thick ascending limb [6–8] and that a corticomedullary ammonium gradient is present in mammalian kidneys [15, 16], some important questions remain concerning the multiplication process. First, based on the idea that the ammonium actively reabsorbed from the medullary thick ascending limb must be secreted into the descending limb to allow ammonium recycling around the loop of Henle, we ask the question: Is the thin descending limb of Henle's loop in the outer medulla ($LDL_{om}$) sufficiently permeable to $NH_4^+$ to allow the recycling to occur? Second, recognizing that the inner medullary part of the ascending limb (ATL) does not actively transport $NH_4^+$, we ask the question: Does $NH_4^+$ passively exit from the inner medullary parts of the long loop of Henle to account for $NH_4^+$ accumulation in the inner medullary interstitium? Finally, recognizing that total ammonia transport can occur via $NH_3$ fluxes as well as via $NH_4^+$ fluxes, we ask the question: Are $NH_3$ permeabilities in loop of Henle segments sufficient to allow physiologically significant $NH_3$ transport?

### *$NH_4^+$ Permeability*

$NH_4^+$ permeability coefficients measured in loop of Henle segments are presented in table 1. The fluxes associated with these permeabilities in vivo are dependent on the magnitude of associated electrochemical driving forces (i.e. electrical potential and concentration gradients), and the extent of equilibration of $NH_4^+$ concentrations across long loop epithelia is dependent on the luminal flow rate. However, a reasonable rule of thumb that can be used to interpret these data is that $NH_4^+$ permeabilities above $50 \times 10^{-5}$ cm/s predict rapid $NH_4^+$ fluxes that can result in near equilibration of $NH_4^+$ concentrations across epithelia at normal loop of Henle flow rates in rodents in vivo [12]; values in the range $6 \times 10^{-5}$ to $50 \times 10^{-4}$ cm/s predict low to moderate $NH_4^+$ fluxes, and values below $5 \times 10^{-5}$ cm/s predict small $NH_4^+$ fluxes that may have limited physiological significance.

*Table 1.* $NH_4^+$ permeability coefficients for loop of Henle segments

| Tubule segment | Animal species | $NH_4^+$ permeability $\times 10^{-5}$ cm/s | Reference |
|---|---|---|---|
| PST | Rabbit | 4.5 ± 0.6 | Garvin et al., 1987 [9] |
| $LDL_{om}$ | Chinchilla | 72 ± 10 | Flessner et al., 1993 [11] |
| $LDL_{im}$ | Chinchilla | 2 ± 19[1] | Flessner et al., 1993 [11] |
| ATL | Chinchilla | 171 ± 59 | Flessner and Knepper, 1993 [12] |
| | Rat | 224 ± 30 | Flessner and Knepper, 1993 [12] |
| MTAL | Rabbit | 15 ± 2.5 | Garvin et al., 1988 [7] |
| | Rat | 6.0 ± 0.8 | Good, 1988 [8] |
| | Rat | 17 ± 3 | Flessner and Knepper, 1993 [12] |

[1] Not significantly different from zero; method was capable of detecting values of $5 \times 10^{-5}$ cm/s or larger.

Among the descending limb segments (i.e. PST; $LDL_{om}$ and $LDL_{im}$), only the $LDL_{om}$ had a high $NH_4^+$ permeability (table 1). Mathematical modelling studies have demonstrated that the $NH_4^+$ permeability in the $LDL_{om}$ is sufficient to allow large $NH_4^+$ fluxes into the lumen with realistic electrochemical gradients [17]. Thus, we conclude that $NH_4^+$ can passively enter the $LDL_{om}$ after having been actively absorbed from the MTAL, thus completing the $NH_4^+$-recycling between the limbs of the loop of Henle. This result, therefore, adds to the evidence supporting the presence of countercurrent multiplication of ammonium in the outer medulla. The $NH_4^+$ permeability values in the PST and $LDL_{im}$ predict $NH_4^+$ fluxes that may be too low to be physiologically significant.

In contrast to the $LDL_{im}$, the ATL has an extraordinarily high $NH_4^+$ permeability (table 1), comparable to the high permeabilities of other monovalent cations in the ATL [18]. This result answers the second question posed above: The high $NH_4^+$ permeability of the ATL provides a pathway for passive $NH_4^+$ absorption that can provide a source of $NH_4^+$ for accumulation in the inner medulla. The measured permeability is sufficient to allow virtually complete equilibration of $NH_4^+$ across the ATL at physiological flow rates [12].

The $NH_4^+$ permeability of the MTAL has been reported to be in the range $6–17 \times 10^{-5}$ cm/s (table 1). Mathematical modelling studies [19] have demonstrated that this permeability is sufficient to account for significant voltage-driven passive absorption of $NH_4^+$ from the MTAL at physiological transepithelial potential differences. Voltage-driven passive $NH_4^+$ transport is thought to account for about one-third of the net absorption of total ammonia meas-

*Table 2.* $NH_3$ permeability coefficients for loop of Henle segments

| Tubule segment | Animal species | $NH_3$ permeability µm/s | Reference |
|---|---|---|---|
| PST | Rabbit | 150±40 | Garvin et al., 1987 [9] |
| $LDL_{om}$ | Chinchilla | 214±40 | Flessner et al., 1993 [11] |
| $LDL_{im}$ | Chinchilla | 287±113 | Flessner et al., 1993 [11] |
| ATL | Chinchilla | 128±93 | Flessner and Knepper, 1993 [12] |
| | Rat | 339±98 | Flessner and Knepper, 1993 [12] |
| MTAL | Rabbit | 31±5 | Garvin et al., 1988 [7] |
| | Rat | 3±5[1] | Flessner and Knepper, 1993 [12] |

[1] Not significantly different from zero; method was capable of detecting values ≥ 30 µm/s.

ured in isolated perfused MTAL segments [7, 8]. The other two-thirds is due to active transport.

*$NH_3$ Permeability*

$NH_3$ permeability coefficients measured in loop of Henle segments are presented in table 2. The fluxes associated with these permeabilities in vivo are dependent on the magnitude of the $NH_3$ concentration gradients across the epithelia. Furthermore, the extent of equilibration of $NH_3$ across the epithelium depends on the luminal flow rate. However, a helpful rule of thumb that can be used to interpret these values (in the context of normal flow rates in rodent long loops of Henle) is that $NH_3$ permeabilities above 500 µm/s predict rapid $NH_3$ fluxes which can cause near equilibration of $NH_3$ concentrations across the epithelium, values in the range 50–500 µm/s predict moderate $NH_3$ fluxes may be physiologically significant but do not result in equilibration at usual in vivo flow rates for the loop of Henle, and values below 50 µm/s predict very low $NH_3$ transport rates.

From the $NH_3$ permeability coefficients reported in table 2, two general observations can be made. First, the $NH_3$ permeability of the MTAL is substantially lower than that of the other segments. As discussed previously [7], the low $NH_3$ permeability of the MTAL is critical to the ability of the MTAL to lower its luminal $NH_4^+$ concentration and thus is critical to its ability to drive countercurrent multiplication of ammonium. Without the low $NH_3$ permeability, active $NH_4^+$ absorption would be negated by rapid $NH_3$ back-fluxes. Second, the $NH_3$ permeability coefficients of the segments other than the MTAL are high enough to permit physiologically significant $NH_3$ fluxes but not so high as to suggest complete equilibration of $NH_3$ concentrations under in vivo conditions. These results, therefore, run counter to the long held concept that plasma membranes possess extraordinarily high permeabilites to $NH_3$ such that $NH_3$ would be in diffusion equilibrium throughout the kidney.

## References


1 Knepper MA, Rector FC Jr: Urinary concentration and dilution; in Brenner BM, Rector FC Jr (eds): The Kidney. Philadelphia, Saunders, 1991, vol 1, pp 445–482.

2 Knepper MA, Packer R, Good DW: Ammonium transport in the kidney. Physiol Rev 1989;69: 179–249.

3 Good DW, Knepper MA: Mechanisms of ammonium excretion: Role of the renal medulla. Semin Nephrol 1990;10:166–173.

4 Knepper MA: $NH_4^+$ transport in the kidney. Kidney Int 1991;40:S95–S102.

5 Knepper MA, Flessner MF: Renal acid-base transport (chapt 6); in Schrier RW, Gottschalk CW (eds): Diseases of the Kidney. Boston, Little Brown, 1993, vol 1, pp 207–232.

6 Good DW, Knepper MA, Burg MB: Ammonia and bicarbonate transport by thick ascending limb of rat kidney. Am J Physiol 1984;247:F35–F44.

7 Garvin JL, Burg MB, Knepper MA: Active $NH_4^+$ absorption by the thick ascending limb. Am J Physiol 1988;255:F57–F65.

8 Good DW: Active absorption of $NH_4^+$ by rat medullary thick ascending limb: Inhibition by potassium. Am J Physiol 1988;255:F78–F87.

9 Garvin JL, Burg MB, Knepper MA: $NH_3$ and $NH_4^+$ transport by rabbit renal proximal straight tubules. Am J Physiol 1987;252:F232–F239.

10 Flessner MF, Knepper MA: Ammonium transport in collecting ducts. Miner Electrolyte Metab 1990;16:299–307.

11 Flessner MF; Mejia R, Knepper MA: Ammonium and bicarbonate transport in isolated perfused rodent long-loop thin descending limb. Am J Physiol 1993;264:F388–F396.

12 Flessner MF, Knepper MA: Ammonium and bicarbonate transport in isolated perfused ascending limbs of the loop of Henle. Am J Physiol 1993;264:F837–F844.

13 Knepper MA: Ultramicroanalysis of tubular fluid by continuous-flow methodology; Hatano M (ed): Nephrology. Tokyo, Springer, 1991, pp 1438–1444.

14 Chou C-L, Knepper MA: In vitro perfusion of chinchilla thin limb segments: Segmentation and osmotic water permeability. Am J Physiol 1992;263:F417–F426.

15 Packer RK, Desai SS, Hornbuckle K, Knepper MA: Role of countercurrent multiplication in renal ammonium handling: Regulation of medullary ammonium accumulation. J Am Soc Nephrol 1991;2:77–83.

16 Knepper MA, Desai SS, Hornbuckle K, Packer RK: Regulation of renal medullary ammonium accumulation. Contrib. Nephrol. Basel, Karger, 1991, vol 92, pp 119–123.

17 Mejia R, Flessner MF, Knepper MA: Model of ammonium and bicarbonate transport along the long-loop thin descending limb: Implications for alkalinization of luminal fluid. Am J Physiol 1993;264:F397–F403.

18 Imai M, Taniguchi J, Tabei K: Function of thin loops of Henle. Kidney Int 1987;31:565–579.

19 Mejia R, Knepper MA: Acid-base transport in the renal thick ascending limb. Math Comp Model 1990;14:538–542.



Mark A. Knepper, MD, PhD, Chief, Renal Mechanisms Section,
Laboratory of Kidney and Electrolyte Metabolism, Building 10,
Room 6N307, National Institutes of Health, Bethesda, MD 20892 (USA)

Tizianello A, Baverel G, Endou H, Schoolwerth AC, O'Donovan DJ (eds):
Renal Ammoniagenesis and Interorgan Cooperation in Acid-Base Homeostasis.
Contrib Nephrol. Basel, Karger, 1994, vol 110, pp 60–66

# Ammonia Transport across Distal Tubule Cells

*K.S. Hering-Smith*[a], *K. Kovach*[b], *L.L. Hamm*[a]

[a] Tulane University School of Medicine, New Orleans, La.;
[b] Washington University School of Medicine, St. Louis, Mo., USA

Until the last decade, total ammonia transport across renal cell membranes was thought to occur via $NH_3$ diffusion only. However, within the last decade, a number of membrane transporters in a variety of nephron segments have been described to transport $NH_4^+$; these include the following: $Na^+$-$H^+$ exchanger, $Na^+$-$K^+$-$2Cl^-$ co-transporter, $Na^+$-$K^+$-ATPase, and $K^+$ channels [1, 2]. Despite these findings, $NH_3$ transport is an important, and incompletely understood, mode of transport in the kidney. In the proximal tubule, significant transport of $NH_4^+$ occurs, but studies with our collaborators clearly indicate significant $NH_3$ diffusion in the proximal tubule [3]. As for the cortical and medullary collecting tubule, our own studies have indicated that the transmembrane permeability to $NH_3$ is significantly less than in the proximal tubule [4, 5]. Despite this restricted diffusion of $NH_3$ in the collecting duct, no mechanism other than $NH_3$ diffusion has been shown to account for transepithelial total ammonia transport in the collecting duct. Knepper et al. [2] have studied a variety of nephron segments and have demonstrated a progressive fall in transepithelial $NH_3$ permeability from the proximal convoluted tubule, to the proximal straight tubule, and to the thick ascending limb and collecting duct. Recently, we have begun using two approaches to address the importance and potential mechanisms of the decreased permeability to $NH_3$ in the distal nephron. First, we have used mathematical modeling of a polarized epithelium to examine the mechanisms and consequences of ammonia transport in renal cells. Secondly, we have used cultured A6 cells as a model epithelium for the distal nephron to examine the mechanism of the low apical membrane permeability to $NH_3$ found in distal tubule cells.

Mathematical modeling of $NH_3/NH_4^+$ transport in a polarized epithelium was undertaken in part to examine the theoretical effects of varying permeabilities to $NH_3$ and $NH_4^+$ on intracellular pH. With most cells, exposure to solutions containing $NH_4Cl$ results in an initial rapid intracellular alkalinization, followed by a slow acidification to (or slightly below) the initial intracellular pH. The initial alkalinization is attributed to the more rapid diffusion of $NH_3$ into the cell. The subsequent slower cell acidification is related to the influx of $NH_4^+$, initially ascribed to movement of $NH_4^+$ down an electrochemical gradient across the cell membrane, related in part to the negative intracellular voltage. (In such cells, subsequent removal of the external $NH_4Cl$ results in a prompt cell acidification. This method of inducing cell acidification has now been widely used to study mechanisms of $H^+$ extrusion in a wide variety of cells.) However, in contrast to the initial alkalinization in most cell types, recently Kikeri et al. [6] demonstrated that luminal introduction of $NH_4Cl$ into the thick ascending limb resulted in cell acidification. This cell acidification was interpreted as due to (a) $NH_4^+$ entry into the cell across the luminal membrane, and (b) a virtual impermeability of the apical membrane to $NH_3$. However, because of the multitude of unknown parameters in their system, other interpretations of the initial cell acidification are feasible. Our goal with mathematical modeling was (in part) to explore the possible factors in the initial acidification in these cells, and to more generally understand the potential effects of varying permeabilities to $NH_3$ and entry of $NH_4^+$ on intracellular pH in a polarized epithelium.

In regard to the mechanism of $NH_3$ transport across cell membranes, several possibilities have been postulated: (1) lipid solubility diffusion; (2) specific transport channels or carriers, and (3) nonspecific small molecule pathways. Lipid solubility diffusion has usually been considered the mechanism whereby $NH_3$ crosses cell membranes which are composed of lipid bilayers. Indeed, Star and Knepper [7] have demonstrated that the transepithelial $NH_3$ permeability in various rabbit nephron segments corresponds with the membrane amplification of each nephron segment, implying that the lipid permeability properties are similar in each nephron segment, the segments just differing in the relative amount of membrane surface area. However, some suggestions have been made that $NH_3$ may cross cell membranes by other mechanisms, not solely by lipid solubility diffusion. For instance, some investigators have noted the correlation between low water permeability and low $NH_3$ permeability in nephron segments such as the thick ascending limb and collecting duct [6]. The corollary to this is that $NH_3$ may cross cell membranes via specific transport channels or carriers, perhaps even water channels. That $NH_3$ may cross cell membranes via nonspecific small molecule pathways can be postulated based on lipid bilayer studies which suggest that the permeation of $NH_3$ through lipid bilayers occurs to a greater extent than explicable by Overton's rule [8]. Overton's rule states that the permeability coefficients of solutes correlate with their oil/water partition coefficients. Some bilayer studies suggest that $NH_3$ (based on its small size) may have some additional component of its passive permeability which is not clearly related to lipid solubility [8]. On the other hand, in some cell membranes, $NH_3$ diffusion has been related to lipid solubility diffusion [9]. However, in cells with a greatly restricted diffusion of $NH_3$, a significant portion of the $NH_3$ permeation might occur by pathways other than lipid solubility diffusion. To address the mechanism(s) of $NH_3$ transport in distal cells, we have begun to use cell culture models.

## Methods

*Mathematical Modeling*

Our simplified model is an expanded version of that originally described by Boron and DeWeer [10]. Basically, their model describes a nonpolarized cell with a fixed intracellular buffering capacity which has a finite and defined permeability to both $NH_3$ and $NH_4^+$. In their model, a fixed negative intracellular voltage drives some influx of $NH_4^+$. Their model was used to demonstrate the time course of changes in intracellular pH in theoretical nonpolarized cells after adding $NH_4Cl$ into the extracellular solution. Numerical solution of a small series of differential equations was used for this purpose. Our simplified model is exactly analogous except that instead of a nonpolarized cell, we used a polarized cell with separate apical and basolateral solutions, and separate defined permeabilities to $NH_3$ and $NH_4^+$ in the apical and basolateral membranes. In this manner, $NH_4Cl$ could theoretically be added to either the apical or basolateral membrane (or both) and the effect on intracellular pH modeled.

*Cell Culture Studies*

A6 cells were grown to confluence on plastic supports in standard tissue culture media. For some experiments, cells were grown on, and studied on, transparent permeable supports so that solution changes could be made separately in the apical and basolateral solutions. Cells were studied at room temperature on an inverted epifluorescent microscope as previously described [11]. Intracellular pH in these cells was measured using the fluorescent dye BCECF, also as previously detailed [11]. To measure rapid changes in intracellular pH, only the single, pH-sensitive excitation wavelength (500 nm) was measured during the periods of rapid changes.

## Results and Discussion

*Mathematical Modeling*

One purpose of our use of mathematical modeling was to address the possible contributing mechanisms to the relatively unique cell acidification seen in the thick ascending limb with luminal $NH_4Cl$ [6]. With the mathematical modeling, two interrelated questions were posed: (1) Does basolateral efflux of $NH_3$>>$NH_4^+$ (rather than only $NH_4^+$ entry across the apical membrane) contribute to the cell acidification? (2) Does cell acidification on luminal addition of $NH_4Cl$ depend upon a low apical membrane permeability to $NH_3$? Realistic parameters for the thick ascending limb for cell volume, surface area, buffering capacity, etc. were assigned, based on values in the literature. With this model and the parameters defined below, cell acidification was simulated with apical addition of $NH_4Cl$. To address the first question, the apical membrane permeability to $NH_3$ and $NH_4^+$ were fixed at $10^{-3}$ cm/s and $5 \times 10^{-3}$ cm/s, respectively. Although these values are undoubtedly high, these values will result in significant entry of both species ($NH_3$ and $NH_4^+$) across the apical membrane on application of $NH_4Cl$ into the lumen. (In fact, $NH_4^+$ undoubtedly enters via $K^+$ channels and the $Na^+$-$K^+$-$2Cl^-$ cotransporter, but a high permeability coefficient will have the same effect to result in significant $NH_4^+$ entry.) With constant apical permeabilities, in separate simulations the basolateral permeabilities to $NH_4^+$ and $NH_3$ were varied. These simulations demonstrated that variations in basolateral permeability to $NH_4^+$ (varying between $10^{-2}$ and $10^{-5}$ cm/s) had relatively little effect on the cell acidification seen with luminal addition of $NH_4Cl$; however, cell acidification was very sensitive to the magnitude of the basolateral permeability to $NH_3$. This implies that the basolateral efflux of $NH_3$>>$NH_4^+$ is a significant contributing factor to the acidification. In fact, $NH_4^+$ entry per se (in isolation) is expected to have little impact on cell pH. This is expected since the pH of a solution buffered at physiologic pH is affected relatively little by the addition of $NH_4Cl$; the release of small amounts of protons to yield small amounts of $NH_3$ ($NH_4^+ \rightarrow NH_3 + H^+$) results in the equilibration of pH only slightly below the starting pH since physiologic pH is nearly 2 orders of magnitude below the pKa of $NH_4^+$ (pKa = approximately 9.0).

To address the second question regarding the importance of the apical membrane permeability to $NH_3$, simulations were performed with fixed basolateral permeabilities to $NH_4^+$ and $NH_3$ ($10^{-5}$ and $5 \times 10^{-2}$ cm/s, respectively). Then, simulations were run with variations in $NH_4^+$ permeability in the apical membrane from $10^{-3}$ to $10^{-5}$ cm/s and apical $NH_3$ permeability from $5 \times 10^{-2}$ to $5 \times 10^{-4}$ cm/s. The qualitative results were that the pattern of cell pH changes was similar over the entire range of apical $NH_3$ permeabilities. Even with an apical membrane permeability to $NH_3$ as high as $5 \times 10^{-2}$ cm/s, there was rapid intracellular acidification seen as long as the apical membrane influx of $NH_4^+$ was adequate. Therefore, these results imply that the cell acidification does depend upon the apical entry of $NH_4^+$ coupled to the efflux of $NH_3$, but *does not* absolutely depend on an apical membrane which is relatively impermeable to $NH_3$. The two determinants of cell acidification in the thick ascending limb appear to be the apical entry of $NH_4^+$ (even with significant $NH_3$ entry) and basolateral efflux of $NH_3$>>$NH_4^+$. An extremely low permeability to $NH_3$ across the apical membrane is not required to result in intracellular acidification. Therefore, these applications of simple mathematical modeling of the effect of $NH_4Cl$ on intracellular pH in the thick ascending limb illustrate the utility of modeling to better understand the mechanisms of ammonia transport; mathematical modeling is particularly useful because of the interrelationship of the two molecular species $NH_3$ and $NH_4^+$ and the differing properties of the two membranes.

*Cell Culture Studies*

The second area of studies involved the use of cultured cells to investigate the potential mechanisms of the decreased permeability to $NH_3$ found in the distal nephron, including both the thick ascending limb and the collecting duct. For the distal nephron we have initially used A6 cells, a model epithelium for the collecting duct which was originally derived from toad kidney. Two types of initial studies suggested that the apical membranes of A6 cells have a relatively low permeability to $NH_3$. First, A6 cells grown on permeable supports did not alkalinize on apical addition of $NH_4Cl$, in contrast to abrupt alkalinization of intracellular pH on basolateral addition of $NH_4Cl$. Second, in A6 cells grown on plastic, addition of $NH_4Cl$ into the media caused an intracellular alkalinization, but at a slower rate than in primary cultures of rabbit proximal tubule cells or inner medullary collecting duct cells (half-times of alkalinization were $4.9 \pm 0.6$, $1.0 \pm 0.1$, and $0.9 \pm 0.1$ s, respectively, for the three cell types). These studies demonstrate that the A6 cells have a low apical membrane permeability to $NH_3$. Other control experiments demonstrated that inhibition of $Na^+$-$H^+$ exchange did not alter the cell alkalinization in A6 cells; this is important since $NH_4^+$ exchange with intracellular $H^+$ can cause cell alkalinization in some cells [12].

To gain insight into the potential mechanisms of the low $NH_3$ permeability in A6 cells, we have used a variety of low molecular weight organic acids to compare the relative permeability (as a function of lipid partition coefficient) to $NH_3$. The question being posed is whether the permeability of $NH_3$ is predictable based solely on its known lipid solubility, or does some other factor (e.g., molecular size, nature of the molecule, etc.) make its permeability different from compounds with a similar degree of lipid solubility. A series of carboxylic acids (formic acid, acetic acid, propionic acid, butyric acid, and valeric acid) was used to evaluate relative permeabilities. Use of these acids has several advantages: the permeabilities of the unionized acid moiety can be estimated from changes in cell pH; the lipid partition coefficients are known, and their size is relatively small although not as small as $NH_3$. For the

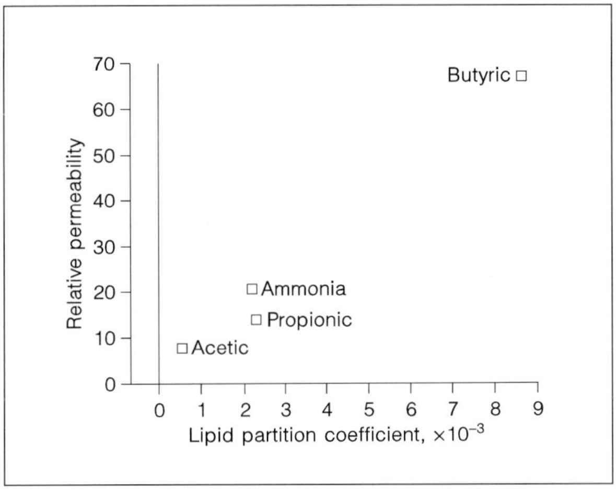


*Fig. 1.* Relative permeabilities of $NH_3$ and certain organic acids as a function of their lipid partition coefficients.

permeability measurements, solutions containing 20 m*M* of $NH_4Cl$ or one of the weak acids was added into the apical solution. Relative permeability coefficients could then be calculated from the changes in cell pH and the measured buffering capacities. These data are displayed in figure 1. The data suggest that $NH_3$ behaves as expected for its known lipid partition coefficient. The apparent low $NH_3$ permeability in these cells is not unique to $NH_3$, but is predictable on the basis of the permeability of unrelated compounds with similar lipid solubility. Similar studies are being extended to other cell types. Therefore, the low $NH_3$ permeability in certain distal nephron cells may be a property of their apical membrane lipid composition, not a function of the presence or absence of particular membrane transporters or proteins.

## References

1 DuBose TF Jr, Good DW, Hamm LL, Wall SM: Ammonium transport in the kidney: New physiological concepts and their clinical applications. J Am Soc Nephrol 1991;1:1193–1203.
2 Knepper MA, Packer R, Good DW: Ammonium transport in the kidney. Physiol Rev 1989;69:179–238.
3 Hamm LL, Simon EE: Ammonia transport in the proximal tubule. Miner Electrolyte Metab 1990;16:283–290.
4 Hamm LL, Trigg D, Martin D, Gillespie C, Buerkert J: Transport of ammonia in the rabbit cortical collecting tubule. J Clin Invest 1985;75:478–485.

5 Hamm LL: Ammonia transport in the rabbit medullary collecting tubule. Kidney Int 1986;29: 367A.
6 Kikeri D, Sun A, Zeidel ML, Hebert SC: Cell membranes impermeable to $NH_3$. Nature 1989; 339:478–480.
7 Star RA, Kurtz I, Mejia R, Burg MB, Knepper MA: Disequilibrium pH and ammonia transport in isolated perfused cortical collecting ducts. Am J Physiol 1987;253:F1232–F1242.
8 Walter A, Gutknecht J: Permeability of small nonelectrolytes through lipid bilayer membranes. J Membr Biol 1986;90:207–217.
9 Golchini K, Kurtz I: $NH_3$ permeation through the apical membrane of MDCK cells is via a lipid pathway. Am J Physiol 1988;255:F135–F141.
10 Boron WF, DeWeer P: Intracellular pH transients in squid axons caused by $CO_2$, $NH_3$, and metabolic inhibitors. J Gen Physiol 1976;67:91–112.
11 Hering-Smith KS, Cragoe EJ, Weiner ID, Hamm LL: Inner medullary collecting duct $Na^+$-$H^+$ exchanger. Am J Physiol 1991;260:C1300–C1307.
12 Preisig PA, Alpern RJ: Pathways for apical and basolateral membrane $NH_3$ and $NH_4^+$ movement in the rat proximal tubule. Am J Physiol 1990;259:F587–F593.

L.L. Hamm, MD, Section of Nephrology (SL 45), Tulane University Medical Center,
1430 Tulane Avenue, New Orleans, LA 70112-2699 (USA)

Tizianello A, Baverel G, Endou H, Schoolwerth AC, O'Donovan DJ (eds):
Renal Ammoniagenesis and Interorgan Cooperation in Acid-Base Homeostasis.
Contrib Nephrol. Basel, Karger, 1994, vol 110, pp 67–74

# In vitro and in vivo Effects of Ammonium Chloride on H-K-ATPase Activity in the Outer Medullary Collecting Duct

*Lal C. Garg, Yasuhiro Komatsu*

Department of Pharmacology and Therapeutics and Division of Nephrology, Hypertension and Transplantation, University of Florida College of Medicine, Gainesville, Fla., USA

The ammonium ion has been shown to substitute for the potassium ion in several transport systems in the plasma membranes. For example, $NH_4^+$ can substitute for $K^+$ in activating Na-K-ATPase [1, 2]. Also, ammonium can be reabsorbed in the thick ascending limb of loop of Henle by replacing $K^+$ on the Na:K:2Cl transporter [3, 4]. Recently, we and other investigators have reported the presence of an ATPase in the mammalian collecting duct segments, which is stimulated by $K^+$ and is not inhibited by ouabain, an inhibitor of Na-K-ATPase [5–7]. This ATPase is inhibited by several inhibitors of gastric H-K-ATPase, such as omeprazole, SCH 28080 and vanadate, suggesting that the renal ouabain-insensitive K-ATPase is pharmacologically similar to the well-characterized gastric H-K-ATPase [7, 8]. The gastric enzyme is stimulated by both KCl and $NH_4Cl$. However, it is not known whether $NH_4Cl$ can also stimulate renal H-K-ATPase.

In addition, in vivo administration of $NH_4Cl$ produces metabolic acidosis which also affects transporters and enzymes responsible for $H^+$ secretion into the renal tubular lumen. H-K-ATPase in the outer medullary collecting duct (OMCD) has been shown to be involved in $H^+$ secretion and $K^+$ reabsorption [9]. The activity of renal H-K-ATPase increases during dietary potassium restriction and is abolished during potassium loading of the animals [10]. However, it is not known whether changes in acid-base balance also influence the renal H-K-ATPase activity.

The purpose of the present study was threefold: (a) to determine if $NH_4Cl$ can substitute for KCl in stimulating H-K-ATPase activity in the OMCD; (b) to determine if H-K-ATPase activity in the OMCD is influenced by metabolic acidosis and alkalosis, and (c) to determine the combined effects of metabolic acidosis and dietary potassium restriction on H-K-ATPase activity in the OMCD.

## Methods

In part one of our study, we measured the effects of $NH_4^+$ on H-K-ATPase activity in the OMCD of rats given a low-potassium diet (0.01% $K^+$, Purina test diet) for 2 weeks. We microdissected the OMCD from the rat kidney by using our previously established method [11], which involves the perfusion of one kidney with chilled Hanks solution containing 0.1% collagenase and incubating several sagittal sections in Hanks solution containing collagenase at 37 °C for 30 min. The OMCD segments from the inner stripe of the outer medulla are then dissected in cold Hanks solution without collagenase under a stereomicroscope.

The length of the microdissected OMCD segment was then measured and several segments were pooled into one sample to get the desired length of the tubule. The total lengths of the pooled OMCD segments was approximately 4–5 mm.

We measured the total ATPase as well as H-K-ATPase activity in the microdissected OMCD segments by using our previously established assay method [5] involving the coupling of ATP hydrolysis with the oxidation of NADH [2]:

$$\text{(1) ATP} \xrightarrow{\text{ATPase}} \text{ADP + Pi}$$

$$\text{(2) ADP + phosphoenolpyruvate} \xrightarrow{\text{pyruvate kinase}} \text{ATP + pyruvate}$$

$$\text{(3) Pyruvate + NADH} \xrightarrow{\text{lactate dehydrogenase}} \text{NAD + lactate}$$

All the reactants in our assay system are in considerable excess; therefore, reactions 2 and 3 are instantaneous. The limiting factor in this assay is reaction 1, the generation of ADP from ATP by cellular ATPases.

The assay procedure is performed in two steps, the first consisting of reactions 1 and 2, and the second of reaction 3. In step one, the microdissected tubules were transferred to 1.5-ml centrifuge tube containing 20 µl of hypotonic solution and kept on ice for at least 15 min. The tubule samples were then frozen in a mixture of dry ice and acetone and kept frozen for at least 30 min. This treatment produces maximum hydrolysis of ATP, probably by increasing the permeability of the tubular cells to ATP [2, 5]. The tubules were then thawed in ice water and 20 µl of incubation buffer was added. The reaction was started by the addition of 20 µl of starting solution to each sample and the incubation of the sample in a shaking water bath kept at 37 °C. The final composition of the incubation mixture was (in m*M*): NaCl, 100; $NH_4Cl$ (or choline chloride), 0–100; imidazole, 50.3; $MgCl_2$, 3.7; EDTA, 0.08; $Na_2ATP$, 1.1; phosphoenolpyruvate, 0.2; ouabain, 4; sodium azide, 2.5, and pyruvate kinase, 2.3 U/ml. The pH of the buffer was 7.4. In our pilot experiments, we observed that the ADP formation increased linearly up to 120 min; we incubated our samples for 40 min. The reactions in this first step were terminated by heating the samples in boiling water for 3 min. The samples were then cooled to room temperature.

In step two, each sample from step one was mixed with 1 ml of 10 μ*M* NADH solution in 0.1 *N* phosphate buffer saline containing 10 U/ml of lactate dehydrogenase. The fluorescence of the unreacted NADH in each sample and in solutions for standard curve (ADP solutions without tubules) was measured in a fluorometer (Perkin-Elmer) set at 345 n*M* for excitation and at 465 n*M* for emission.

H-K-ATPase activity is defined as the difference between the ATPase activity in the presence and in the absence of KCl (or $NH_4Cl$) but in the presence of ouabain and sodium azide. Ouabain is added to inhibit Na-K-ATPase activity and sodium azide is added to inhibit mitochondrial ATPases. In our pilot experiments, we observed that H-K-ATPase activity was proportional to the total length of the OMCD segments (at least up to 8 mm); therefore, we have calculated H-K-ATPase activity as pmol ADP/min/mm tubule.

In part two of our study, we measured H-K-ATPase activity in 6 groups of male Sprague-Dawley rats: (1) the control group of rats were given a regular diet (Ralston Purina, St. Louis, Mo., USA) with free access to tap water for 1 week; (2) the metabolic acidosis group of rats were given a regular diet and a 1.5% $NH_4Cl$ solution to drink for 1 week; (3) the metabolic alkalosis group of rats were given a regular diet and 0.28 *N* $NaHCO_3$ in their drinking water for 1 week; (4) the K-restricted rats were given a specially formulated low-$K^+$ diet (Ralston Purina) for 2 weeks; (5) the K-restricted and acid-loaded rats were given a low-$K^+$ diet for 1 week followed by a low-$K^+$ diet and 1.5% $NH_4Cl$ in their drinking water for 1 week, and (6) the K-restricted and alkali-loaded rats were given a low-$K^+$ diet for 1 week followed by a low-$K^+$ diet and 0.28 *N* $NaHCO_3$ in their drinking water for 1 week. All animals were housed in a room with an ambient temperature of 26 °C and a 12-hour light and dark cycle. The urine of each animal was collected in a metabolic cage (Nalgene) for 24 h prior to experimentation. On the day of the experiment, a blood sample from each animal was collected for chemical analysis. Blood pH and $pCO_2$ were determined by the Blood Gas Analyzer (Instrumentation Laboratory, Lexington, Mass., USA) and blood bicarbonate was calculated by the Henderson-Hasselbalch equation. The concentrations of $Na^+$ and $K^+$ in urine and plasma were measured by using a Na/K Analyzer (Ciba-Corning). The methods used for the microdissection of the OMCD from these 6 groups of rats and for the assay of ATPase were the same as described above.

## Results

$NH_4Cl$ increased ouabain-insensitive ATPase activity in a concentration-dependent manner (fig. 1). The stimulation of ATPase activity was observed at a very low (1 m*M*) concentration of $NH_4Cl$. The effect reached a plateau at approximately 60 m*M* $NH_4Cl$. The $NH_4Cl$-stimulated ATPase activity was inhibited by vanadate (fig. 2), SCH 28020 and omeprazole (data not shown) as has been reported in the case of KCl [5]. These results indicate that $NH_4^+$-stimulated ATPase activity in our system is identical to H-K-ATPase activity.

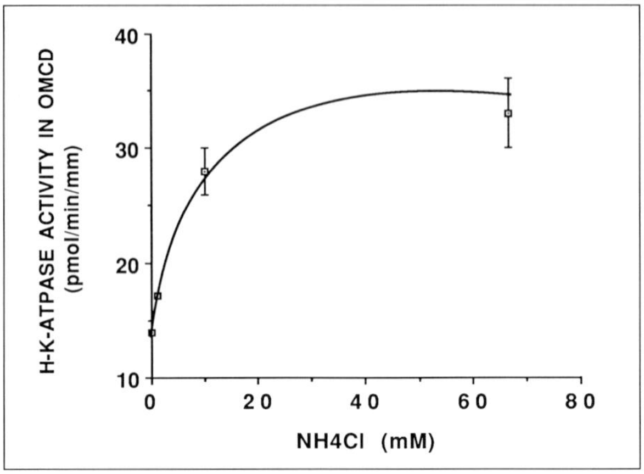


1

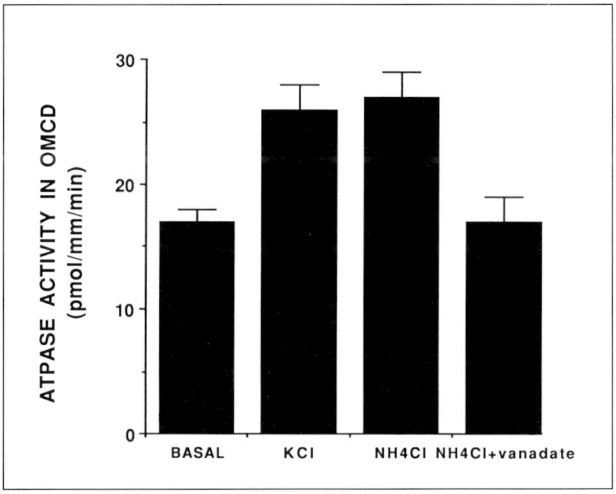


2

*Fig. 1.* Effect of $NH_4Cl$ on ouabain-insensitive ATPase activity in the OMCD of $K^+$-deprived rats. Data points are the mean ± SEM of 3 animals.

*Fig. 2.* Effect of vanadate on ouabain-insensitive and $NH_4Cl$-stimulated ATPase activity in the OMCD of $K^+$-deprived rats. Each bar represents mean ± SEM of 4–6 animals.

Table 1 shows the acid-base status and plasma $K^+$ values of the 6 groups of rats used in the study to determine the effects of acid and alkali load with and without dietary $K^+$ restriction. Plasma bicarbonate and pH were lower in acid-loaded animals than in control or alkali-loaded animals among the 3 groups of rats kept in a regular diet. Acid loading also decreased plasma bicar-

*Table 1.* Plasma $K^+$ and acid-base status of rats

| Treatment | n | $HCO_3^-$, mEq/l | pH | $K^+$, mEq/l |
|---|---|---|---|---|
| Control | 3–4 | $24.87 \pm 0.43$ | $7.33 \pm 0.02$ | $4.14 \pm 0.1$ |
| Acid load | 3–6 | $17.21 \pm 1.02^{a,b}$ | $7.14 \pm 0.04^{a,b}$ | $4.27 \pm 0.64$ |
| Alkali load | 4 | $25.17 \pm 0.86$ | $7.37 \pm 0.01$ | ND |
| Low-K | 6–8 | $23.43 \pm 1.94$ | $7.33 \pm 0.05$ | $3.38 \pm 0.3$ |
| Low-K + acid | 7–14 | $14.62 \pm 1.68^{a,b}$ | $7.14 \pm 0.04^{a,b}$ | $2.67 \pm 0.26^{a}$ |
| Low-K + alkali | 5–8 | $37.05 \pm 3.25^{b}$ | $7.43 \pm 0.04^{a,b}$ | $4.41 \pm 0.33^{b}$ |

Each value is the mean ± SEM of number of animals.
[a] $p < 0.05$ vs. control. [b] $p < 0.05$ vs. low-K alone. ND = Not done because of hemolysis.

*Table 2.* Urine flow and excretion $K^+$ and $NH_4^+$

| Treatment | n | $K^+$ µEq/day | Ammonia µEq/day | Urine flow ml/day |
|---|---|---|---|---|
| Control | 10 | $1,862 \pm 205$ | $2,458 \pm 792$ | $13.5 \pm 0.9$ |
| Acid load | 14 | $1,981 \pm 206$ | $9,233 \pm 1,101^{a}$ | $15.1 \pm 1.7^{b}$ |
| Alkali load | 10 | $1,580 \pm 229$ | $1,766 \pm 714$ | $534.7 \pm 8.9^{a}$ |
| Low-K | 7–9 | $55.8 \pm 23.8^{a}$ | $3,271 \pm 653$ | $30.8 \pm 9.2$ |
| Low-K + acid | 11 | $227 \pm 55.7^{a,b}$ | $5,540 \pm 938^{a}$ | $9.4 \pm 1.4^{a,b}$ |
| Low-K + alkali | 5 | $ND^{a,b}$ | $920 \pm 520^{a,b}$ | $24.4 \pm 3.8^{a}$ |

Each value is the mean ± SEM of number of animals. ND = Not detectable.
[a] $p < 0.05$ vs. control. [b] $p < 0.05$ vs. low-K alone.

bonate and pH in low-$K^+$ animals. In addition, alkali loading increased plasma bicarbonate and pH in low-$K^+$ animals. All 3 groups of low-$K^+$ animals had lower plasma $K^+$ concentrations than did the animals on the regular diet. There was no significant difference in plasma $Na^+$ concentration among the 6 groups of animals (data not shown).

Urinary $K^+$ excretion was decreased from 1,862 µEq/day in the control group of rats to <230 µEq/day in the low-K diet group of animals (table 2). Potassium excretion was lowest (nondetectable) in the low-K diet alkali-loaded group and was highest in the low-K diet acid-loaded group among 3 groups of low-K diet animals. The excretion of ammonia in the acid-load group (9,233 µEq/ day) on regular diet was not only greater than that of control group (2,458 µEq/ day) but was also greater than that of acid-load group on low-K diet (5,540 µEq/day).

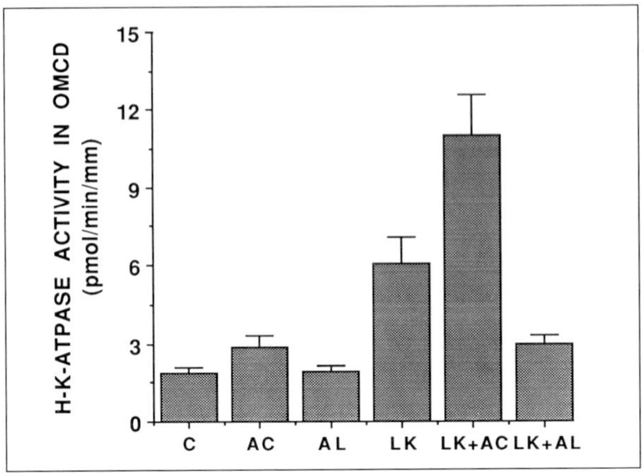


*Fig. 3.* Effects of acid-base changes on H-K-ATPase activity in the OMCD of rats given normal diet (C) and low-K diet (LK). AC = Acidosis; AL = alkalosis; LK = low-K diet. Each bar represents the mean ± SEM of 4–11 rats.

Acid treatment alone increased H-K-ATPase activity in the OMCD from $1.9 \pm 0.2$ to $2.9 \pm 0.4$ pmol/min/mm (fig. 3). The low-$K^+$ diet alone increased the enzyme activity to $6.1 \pm 1.1$ pmol/min/mm, and a combination of low-$K^+$ diet and $NH_4Cl$ increased the enzyme activity further to $11.5 \pm 2.0$ pmol/min/mm. The enzyme activities in the OMCD with $NaHCO_3$ treatment alone and with a combination of low-$K^+$ diet and $NaHCO_3$ were $1.8 \pm 0.3$ and $3.0 \pm 0.4$ pmol/min/mm, respectively.

## Discussion

The present studies provide evidence that $NH_4^+$ stimulates ouabain-insensitive ATPase activity in the rat OMCD. This ATPase is inhibited by vanadate, SCH 28080 and omeprazole, suggesting that $NH_4^+$-stimulated ATPase is pharmacologically similar to the renal H-K-ATPase [5]. The $K_m$ of renal H-K-ATPase for $NH_4Cl$ is approximately 2 m*M* (fig. 1) as compared with <1 m*M* for KCl [5]. The physiological significance of the stimulation of H-K-ATPase activity by $NH_4^+$ in the OMCD is not known at this time. It is possible that H-K-ATPase may transport $NH_4^+$ into the OMCD cell in exchange for $H^+$ when the concentration of $NH_4^+$ is much higher than that of $K^+$ in the tubular lumen (fig. 4). The $NH_4^+$ may then dissociate into $H^+$ and $NH_3$ inside the cell and $NH_3$ may either diffuse into the tubular lumen causing recycling of ammonia or diffuse into the peritubular fluid producing a net reabsorption of

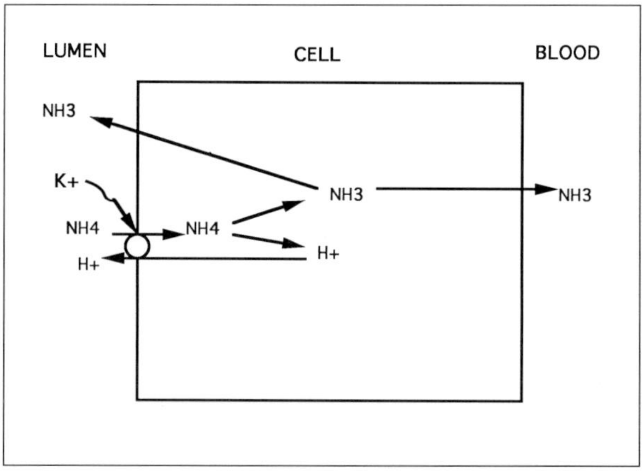


*Fig. 4.* A model for transport of $NH_4^+$ in the OMCD. $NH_4^+$ absorbed by H-K-ATPase dissociates into $H^+$ and $NH_3$. $H^+$ is secreted into the tubular lumen via H-K-ATPase and $NH_3$ may diffuse into the tubular lumen and/or into the blood.

ammonia in the OMCD. The decreased excretion of ammonia in our acid-load group on low-K diet than in acid-load group on regular diet (table 2) suggests that $NH_4^+$ may be reabsorbed in the OMCD when the luminal $K^+$ concentration is decreased. This phenomenon produces a greater urinary excretion of $K^+$ and a lower plasma $K^+$ in animals of low-K plus acid-load group than low-K group (tables 1, 2) in spite of $K^+$ deficit in both groups.

The second major finding of our studies is that the metabolic acidosis stimulates H-K-ATPase activity in the rat OMCD. The changes in acid-base balance influence OMCD H-K-ATPase activity to a much larger extent in low-K rats than in the normal animals.

*In summary,* we have demonstrated that: (a) $NH_4^+$ can substitute for $K^+$ in the stimulation of renal H-K-ATPase in the rat OMCD; (b) H-K-ATPase activity in the OMCD is stimulated by both dietary $K^+$ restriction and metabolic acidosis, and (c) dietary $K^+$ restriction potentiates the effects of acid-base changes on renal H-K-ATPase.

## References

1 Post RL, Jolly PC: Linkage of sodium, potassium, and ammonium active transport across erythrocyte membrane. Biochim Biophys Acta 1957;25:118–128.
2 Garg LC, Knepper MA, Burg MB: Mineralocorticoid effects on Na-K-ATPase in individual nephron segments. Am J Physiol 1981;240:F536–F544.
3 Garvin JL, Burg MB, Knepper MA: Active $NH_4^+$ absorption by the thick ascending limb. Am J Physiol 1988;255:F57–F65.

4 Good DW: Active absorption of $NH_4^+$ by rat medullary thick ascending limb: Inhibition by potassium. Am J Physiol 1988;255:F78–F87.
5 Garg LC, Narang N: Ouabain-insensitive K-ATPase in distal nephron segments of the rabbit. J Clin Invest 1988;81:1204–1208.
6 Doucet A, Marsy S: Characterization of K-ATPase activity in distal nephron: Stimulation by potassium depletion. Am J Physiol 1987;253:F418–F423.
7 Garg LC: Respective roles of H-ATPase and H-K-ATPase in ion transport in the kidney. J Am Soc Nephrol 1991;2:949–960.
8 Sachs G, Carlsson E, Lindberg P, Wallmark B: Gastric H, K-ATPase as therapeutic target. Annu Rev Pharmacol Toxicol 1988;28:269–284.
9 Wingo CS: Active proton secretion and potassium absorption in the rabbit outer medullary collecting duct. J Clin Invest 1989;84:361–365.
10 Garg LC, Narang N: Suppression of ouabain-insensitive K-ATPase activity in rabbit nephron segments during chronic hyperkalemia. Renal Physiol Biochem 1989;12:295–301.
11 Garg LC, Mackie S, Tisher CC: Effect of low potassium diet on Na-K-ATPase in rat nephron segments. Pflügers Arch 1982;394:113–117.

Lal C. Garg, PhD, Department of Pharmacology and Therapeutics, PO Box 100267,
University of Florida Health Sciences Center, Gainesville, FL 32610–0267 (USA)

Tizianello A, Baverel G, Endou H, Schoolwerth AC, O'Donovan DJ (eds):
Renal Ammoniagenesis and Interorgan Cooperation in Acid-Base Homeostasis.
Contrib Nephrol. Basel, Karger, 1994, vol 110, pp 75–80

# Mechanism of $NH_4^+$ Transport in the Rat Collecting Duct

*Susan M. Wall*

Division of Nephrology, Department of Medicine, University of Texas,
Medical School at Houston, Tex., USA

## The Mechanism of $NH_4^+$ Transport

The kidney excretes acid by reclaiming filtered bicarbonate and excreting protons buffered with inorganic anions such as phosphate and ammonia ($NH_3$) [1]. Urinary net acid is the sum of urinary ammonium ($NH_4^+$) plus titratable acid (primarily phosphate) minus urinary bicarbonate. $NH_4^+$ excretion varies appropriately with perturbations in acid-base balance and accounts for the major regulatable portion of net acid excretion [1]. Thus, ammonium excretion is critical for regulation of systemic arterial pH [1–3].

Ammonia is synthesized in the proximal tubule and secreted into the tubule lumen. In the loop of Henle $NH_4^+$ is absorbed and concentrated in the medullary interstitium by countercurrent multiplication. Interstitial ammonium is then secreted into the collecting duct by transepithelial transport. Roughly 80% of urinary ammonium is secreted along the collecting duct [4].

How $NH_4^+$ is transported from the interstitium to the collecting duct lumen is not fully understood. Ammonia has a pK of 9.0 at 37 °C. Therefore, as pH is reduced, progressively more of total ammonia ($NH_4^+$ and $NH_3$) is $NH_4^+$ and less is $NH_3$. Since at physiological pH, most total ammonia is $NH_4^+$, total ammonia is often referred to as ammonium ($NH_4^+$). Ammonium transport could occur by (1) direct transport of $NH_4^+$ or (2) passive $NH_3$ transport in parallel with active proton transport. To date there is little evidence that ammonium secretion across the collecting duct epithelium occurs by direct $NH_4^+$ transfer [1].

The flux of $NH_3$, $J_{NH_3}$, across an epithelium is given by the equation: $J_{NH_3} = A_s P_{NH_3} \Delta C_{NH_3}$, where $A_s$ is the epithelial surface area, $P_{NH_3}$ is the epithelial $NH_3$ permeability, and $\Delta C_{NH_3}$ is the $NH_3$ concentration gradient between the interstitium and the collecting duct lumen. Each of these factors will be considered individually.

*Surface Area*

Knepper et al. [5] and Mejia et al. [6] measured total inner epithelial surface area of rat collecting duct using morphometric techniques. They found a gradual decrease in collecting duct surface area from the proximal to the distal regions. A rapid decrease in surface area occurred in the inner medullary collecting duct (IMCD), due to progressive merging of collecting ducts as one goes deeper into the inner medulla.

*Permeability to $NH_3$ and $NH_4^+$*

Flessner et al. [7] measured $NH_3$ and $NH_4^+$ permeability for each collecting duct segment in the rat using the isolated perfused tubule technique. $NH_4^+$ permeability was low in each segment studied, roughly $10^{-5}$ cm/s. In contrast, an axial gradient along the collecting duct was found for $NH_3$ permeability. Permeability was lowest in the IMCD (0.002 cm/s), highest in the cortical collecting duct (CCD, 0.023 cm/s) and intermediate in the inner stripe of the outer medullary collecting duct (OMCD). High $NH_3$ permeability with negligible $NH_4^+$ permeability enables $NH_3$ secretion with little back diffusion of $NH_4^+$.

*Transepithelial $NH_3$ Concentration Gradient, $\Delta C_{NH_3}$*

The $\Delta C_{NH_3}$ occurs from increased interstitial $NH_3$ concentration generated by countercurrent multiplication and from a decreased luminal $NH_3$ concentration.

*Increased $NH_3$ Concentration in the Interstitium.* Tissue slice studies in the dog [8] and rat [9] showed an axial gradient for interstitial total ammonia concentration. Total ammonia was greatest at the papillary tip and decreased gradually as one approached the inner/outer medullary junction. In the inner medulla total ammonia concentration was increased with acid loading [8, 9] and decreased with alkali loading [9]. Papillary interstitial and collecting duct luminal $NH_3$ concentrations were measured directly in micropuncture studies by Good et al. [10]. Interstitial $NH_3$ concentration was higher than the luminal, thus favoring $NH_3$ secretion. This $NH_3$ concentration difference was accentuated with acidosis, correlating with in vivo observations.

*Decreased Luminal $NH_3$ Concentration.* Ammonium secretion in the collecting duct occurs by active proton secretion in parallel with $NH_3$ diffusion, making luminal proton secretion important in $NH_4^+$ secretion. Luminal acidification along the collecting duct occurs principally through a proton-ATPase [1]. In the collecting duct the greatest rate of proton pumping occurs in the OMCD [11–14].

A fall in luminal pH can be augmented through a disequilibrium pH [1]. This reduction in pH serves to further reduce luminal $NH_3$ concentration. A disequilibrium pH occurs with proton secretion in the absence of luminal carbonic anhydrase activity. Hydrogen ion is secreted into the tubule lumen and rapidly combines with bicarbonate forming carbonic acid. In the presence of luminal carbonic anhydrase, carbonic acid rapidly dehydrates to $CO_2$ and water. Because of the high permeability of the collecting duct epithelium to $CO_2$, the net result is rapid absorption of bicarbonate. In the absence of carbonic anhydrase, carbonic acid dehydrates slowly and therefore accumulates. The accumulation of this strong acid lowers the pH further than that predicted by the Henderson-Hasselbalch equation (disequilibrium pH). This disequilibrium pH shifts the buffer equilibrium for ammonia, favoring the formation of $NH_4^+$ relative to $NH_3$. Hence luminal $NH_3$ is lowered and the driving force, or concentration gradient for $NH_3$ between the interstitium and the collecting duct lumen, is greater.

Carbonic anhydrase is present in the apical membrane of some nephron segments [1]. We profiled the rat collecting duct for carbonic anhydrase activity [13, 15] and found a disequilibrium pH in the CCD and inner stripe of the OMCD. However, endogenous luminal carbonic anhydrase activity was found in the initial IMCD, but not in the terminal IMCD. Studies in the rabbit [16, 17] showed a similar pattern. Our results in the rat terminal IMCD confirmed early micropuncture and microcatheterization [1] studies which demonstrated a luminal disequilibrium pH.

The purpose of the present study was to determine a model for $NH_3$ secretion, generating hypothetical rates of cumulative $NH_3$ secretion rates along the rat collecting duct. Permeabilities for $NH_3$ and $NH_4^+$ [7] and inner epithelial surface areas previously measured for each collecting duct segments [5, 6] were incorporated. In the inner medulla the $\Delta C_{NH_3}$ previously measured [10] was used. In the cortex and outer medulla the $\Delta C_{NH_3}$ was hypothesized based on tissue slice studies which measured total tissue $NH_3$ at a given level [8, 9].

## Results and Discussion

The previous data were combined to give an estimate of the relative contribution of each collecting duct segment to total ammonia excretion. The solid line on figure 1 shows the product of $P_{NH_3}$ and collecting duct surface area. This product is greatest in the CCD and declines gradually as the collect-

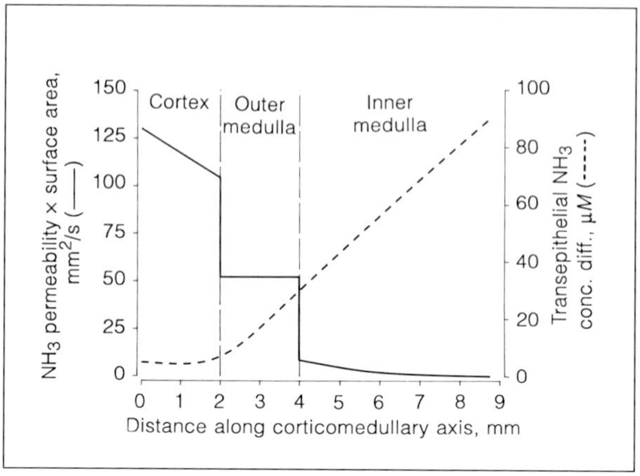


1

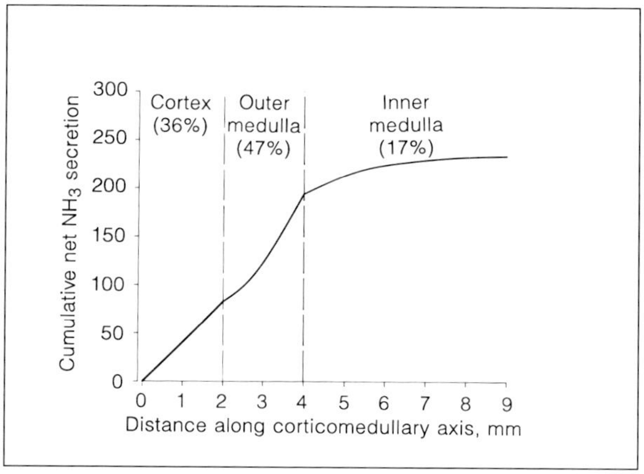


2

*Fig. 1.* The solid line represents the product of total inner epithelial surface area [5, 6] at various levels along the collecting duct and $NH_3$ permeability for each collecting duct segment [7, 15]. The dashed line represents the concentration gradient for $NH_3$ between the interstitium and the collecting duct lumen. This value has been measured in the IMCD [10] and is inferred for the OMCD and CCD [8, 9].

*Fig. 2.* $NH_3$ flux is hypothesized for each collecting duct segment based on the product of $A_s$, $P_{NH_3}$ and $\Delta C_{NH_3}$.

ing duct approaches the outer medulla. Values continue to decline as one approaches the papillary tip. The dashed line shows the hypothetical $NH_3$ concentration gradient.

The $\Delta C_{NH_3}$ increases toward the papillary tip due to countercurrent multiplication. The $\Delta C_{NH_3}$ is highest where the product of surface area and permeability is lowest. Similarly, where this product is highest, $\Delta C_{NH_3}$ is lowest. Seemingly these two factors compensate for one another.

Ammonium flux (fig. 2) was determined by multiplying surface area, $NH_3$ permeability and the hypothetical $NH_3$ concentration gradient. Using this formula we calculated the hypothetical rates of ammonium secretion in each collecting duct segment. These considerations predict that the bulk of ammonium secretion occurs in the CCD and in the OMCD with relatively less ammonium secretion in the IMCD.

This model agrees with the microcatheterization [4] studies which showed ammonium secretion is maintained at a relatively uniform rate along the length of the collecting duct, supporting the conclusion that the rate of secretion of ammonium is not due to $\Delta C_{NH_3}$ alone. Factors which determine collecting duct ammonium secretion can be summarized as follows: In the CCD, as in the OMCD, $\Delta C_{NH_3}$ is relatively low. Therefore, ammonium secretion occurs due to its large surface area, its high $NH_3$ permeability and the presence of a disequilibrium pH. In the OMCD the $NH_3$ gradient from interstitium to lumen is unknown, but may be considerably lower than in the inner medulla. The disequilibrium pH might be highest in this segment because it generates the greatest rate of proton secretion [11–14], thus augmenting net ammonium secretion. A reduction in luminal pH decreases luminal $NH_3$ concentration. The driving force for $NH_3$ flux into the lumen is determined both by the $NH_3$ concentration within the tubule lumen as well as by accumulation of $NH_3$ in the interstitium [1]. Therefore, a reduction in luminal pH indirectly increases the driving force for $NH_3$ secretion. In the terminal IMCD, $NH_3$ permeability, and surface area are relatively low. Therefore, ammonium secretion is maintained by its large interstitial to luminal $NH_3$ gradient and the disequilibrium pH present. Throughout most of the collecting duct a disequilibrium pH is found [1, 13, 15, 16] promoting $NH_4^+$ flux. Luminal carbonic anhydrase was found in only a short segment of the collecting duct [15, 17]. The complexity of ammonium secretion along the collecting duct provides many sites for its regulation.

## References


1 Knepper MA, Packer R, Good DW: Ammonium transport in the kidney. Physiol Rev 1989;69: 179–249.
2 Lemann J Jr, Lennon EJ, Goodman AD, Litzow JR, Relman AS: The net balance of acid in subjects given large loads of acid or alkali. J Clin Invest 1965;44:507–517.
3 Relman AS, Lennon EJ, Lemann J Jr: Endogenous production of fixed acid and the measurement of the net balance of acid in normal subjects. J Clin Invest 1961;40:1621–1630.
4 Sajo IM, Goldstein MB, Sonnenberg H, Stinebaugh BJ, Wilson DR, Halperin ML: Sites of ammonia addition to tubular fluid in rats with chronic metabolic acidosis. Kidney Int 1981;20: 353–358.

5 Knepper MA, Danielson RA, Saidel GM, Post RS: Quantitative analysis of renal medullary anatomy in rats and rabbits. Kidney Int 1977;12:313–323.
6 Mejia R, Sands JM, Stephenson JL, Knepper MA: Renal action of atrial natriuretic factor: A mathematical modelling study. Am J Physiol 1989;257:F1146–F1157.
7 Flessner MF, Wall SM, Knepper MA: Permeabilities of rat collecting duct segments to $NH_3$ and $NH_4^+$. Am J Physiol 1991;260:F264–F272.
8 Robinson RR, Owen EE: Intrarenal distribution of ammonia during diuresis and antidiuresis. Am J Physiol 1965;208:1129–1134.
9 Packer RK, Desai SS, Hornbuckle K, Knepper MA: Role of countercurrent multiplication in renal ammonium handling: Regulation of medullary ammonium accumulation. J Am Soc Nephrol 1991;2:77–83.
10 Good DW, Caflisch CR, DuBose TD Jr: Transepithelial ammonia concentration gradients in inner medulla of the rat. Am J Physiol 1987;252:F491–F500.
11 Tomita K, Pisano JJ, Burg MB, Knepper MA: Effects of vasopressin and bradykinin on anion transport by the rat cortical collecting duct. J Clin Invest 1986;77:136–141.
12 Atkins JL, Burg MB: Bicarbonate transport by isolated perfused rat collecting ducts. Am J Physiol 1985;249:F485–F489.
13 Flessner MF, Wall SM, Knepper MA: Ammonium and bicarbonate transport in rat outer medullary collecting ducts. Am J Physiol 1992;262:F1–F7.
14 Wall SM, Sands JM, Flessner MF, Nonoguchi H, Spring KR, Knepper MA: Net acid transport by isolated perfused inner medullary collecting ducts. Am J Physiol 1990;258:F75–F84.
15 Wall SM, Flessner MF, Knepper MA: Distribution of luminal carbonic anhydrase activity along rat inner medullary collecting duct. Am J Physiol 1991;260:F738–F748.
16 Star RA, Kurtz I, Mejia R, Burg MB, Knepper MA: Disequilibrium pH and ammonia transport in isolated perfused cortical collecting ducts. Am J Physiol 1987;253:F1232–F1242.
17 Star RA, Burg MB, Knepper MA: Luminal disequilibrium pH and ammonia transport in outer medullary collecting duct. Am J Physiol 1987;252:F1148–F1157.

Susan M. Wall, MD, Division of Nephrology, University of Texas Health Science Center at Houston, 6431 Fannin, MSB 4.148, Houston, TX 77030 (USA)

Tizianello A, Baverel G, Endou H, Schoolwerth AC, O'Donovan DJ (eds):
Renal Ammoniagenesis and Interorgan Cooperation in Acid-Base Homeostasis.
Contrib Nephrol. Basel, Karger, 1994, vol 110, pp 81–87

# An RT-PCR Approach to Study the Regulation of mRNA Levels for Phosphate-Dependent Glutaminase

*Susan R. DiGiovanni, Mark A. Knepper*

Laboratory of Kidney and Electrolyte Metabolism, National Heart Lung and Blood Institute, National Institutes of Health, Bethesda, Md., USA

In the rat, the increase in ammonium excretion after an acid load is associated with an increase in ammonium production through the phosphate-dependent glutaminase (PDG) and glutamate dehydrogenase (GDH) pathways. Many potential sites of regulation have been proposed including increased delivery of glutamine into the mitochondria, removal of ammonium from the cell, activation of ammoniagenic enzymes and enzymes in related pathways which result in increased flux through the PDG and GDH pathways, altered ratio of $NAD^+$/NADH, and ATP turnover rate [1]. The most extensively investigated area has been the adaptation of enzymes involved in the ammoniagenic pathway and related metabolic pathways. In the rat, it is clear adaptations occur in both PDG [2–4] and GDH [5, 6] in response to an acid load. Furthermore, several studies have demonstrated that mRNA levels for PDG, GDH and PEPCK increase after an acid load [7–10]. Although the regulation of ammoniagenic enzyme expression is only part of the overall response to acid-base perturbations, the mechanisms involved may have broad implications for pH-mediated regulation of many enzymes and transport proteins. Consequently, we are studying regulation of mRNA levels for PDG and GDH.

Because studies by Good and Burg [11] and Good and DuBose [12] clearly indicate that the early portion of the proximal tubule ($S_1$ proximal tubule) is the most important site of regulation of ammonia synthesis and secretion, our studies are focused on this segment. It was difficult, however, to assess the time course of the adaptation of these enzymes from the current literature because we were not able to directly compare studies which used different methods of administering an acid load and different tissue preparations. Therefore, before investigating regulation of mRNA levels, we measured am-

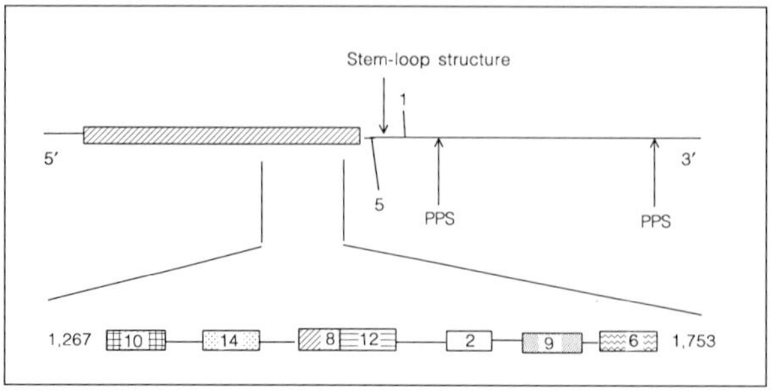


*Fig. 1.* Schematic diagram of PDG cDNA. The large hatched box represents the coding region and the solid lines the untranslated regions. The numbers on the cDNA and the boxes in the expanded portion represent relative positions of primers tested for the PCR reaction. Odd numbers are antisense primers and even numbers are sense primers. The optimum primers were found to be Nos 9 and 10. PPS = Putative polyadenylation signal.

monium production, PDG activity, and GDH activity in $S_1$ proximal tubules from rats undergoing a uniform protocol. These studies demonstrated that the rise in ammonium production by the $S_1$ segment is paralleled by a rise in activity of PDG and not of GDH during the first 4 days of an acid-loaded diet [DiGiovanni et al., unpubl. observations]. We therefore chose to study the regulation of mRNA levels for PDG in the $S_1$ segment.

The cDNA for PDG was cloned in 1991 by Shapiro et al. [13] from a rat brain library. Figure 1 shows a schematic diagram of the cDNA. The coding region is 1,022 bp long and encodes 674 amino acids. The 5' end has a putative mitochondrial targeting sequence. It has a very interesting structure in that the 3' untranslated region (UTR) of the molecule is very long, approximately 800 bases to one potential polyadenylation site and 2.2 kb to the other. There are also multiple inverted repeats which potentially can result in stem-loop structures.

## RT-PCR Technique

Although several techniques are presently available to study mRNA levels, only one, the polymerase chain reaction coupled to reverse transcription (RT-PCR) is sensitive enough to detect changes in mRNA levels in a single microdissected tubule. Therefore, we chose this approach to investigate PDG mRNA levels in rat $S_1$ segments. Two methods of performing RT-PCR in single tubules were developed simultaneously in two different laboratories [14, 15]. In the Briggs [14] approach, tubules are dissected, 10–20 mm of tubule are pooled and the total RNA from the sample is isolated. RT-PCR is then performed with standard procedures (see below). The other approach was developed by García-Pérez and co-workers in our laboratory [15]. In this method, single tubules 1–2 mm in length are dissected, transferred, permeabilized and then RT-PCR is carried out in the same tube. A systematic study comparing the sensitivity of the two approaches has not been done. However the latter approach has the advantage of being simpler, requires less starting material, and since it is all carried out in one tube, the variability associated with multiple manipulations is avoided. Therefore, this is the technique we chose to use.

### Tissue Preparation

The method of Moriyama et al. [15] was modified slightly for our purposes. Except where there are differences, details will not be presented. For initial optimization of the RT-PCR procedure (see below), rats were maintained on a standard rat chow and ad lib drinking water. All solutions and materials when possible are maintained ribonuclease-free. Proximal tubules, identified by attachment to the glomerulus, are dissected from collagenase-treated tissue in the presence of vanadyl ribonucleoside complex (VRC), a potent ribonuclease inhibitor. When all the tubules have been dissected, they are detached from their glomeruli and transferred to a wash dish to rinse them free of VRC. This step is essential as VRC inhibits reverse transcriptase. In the original method described, the tubules were then attached to glass beads and transferred into GeneAmp PCR tubes without carryover of fluid. In our hands, however, proximal tubules do not stay attached to the beads and thus loss of tubules occurs frequently. The tubules are instead transferred with 2 μl of wash solution directly into a 2% Triton X-100 solution containing placental RNase inhibitor and 1 m*M* dithiothreitol. The tubes are then placed on dry ice until all the tubules have been transferred and are ready for reverse transcription. Multiple control studies showed that this method of transfer and permeabilization was effective and yielded comparable results to the previous method with the exception that there is now only rare tubule loss.

### Reverse Transcription

Reverse transcription (RT) makes a single-stranded DNA copy of RNA sequences. There are three effective methods to prime the mRNA for RT; random hexamers, specific primers for the target to be amplified, and oligo(dT). Random hexamers are short oligonucleotides with multiple base combinations represented. This will result in variable length DNA fragments, including ribosomal and transfer RNA. This tends to produce a very high background of nonspecific amplification. Specific primers used for RT can be either the same antisense primer to be used in the PCR or another antisense primer 3' to PCR primer. This has the advantage of reducing background, but in our hands does not appear to be as efficient in RT as other primers. In addition, specific primers do not allow the same cDNA to be used for control amplifications of so-called housekeeping genes or other targets. Oligo(dT) is the most frequently used RT primer. It will result in the RT of all polyadenylated sequences. In general, we try all three primers for every template to be amplified and choose the primer which consistently gives the greatest yield. This has almost always been oligo(dT). There is a limit to the length of cDNA that can be produced by RT and because the glutaminase cDNA (fig. 1) has a very long 3' UTR, we were initially concerned that oligo(dT) would not be an appropriate primer for glutaminase. We have found, however, that it actually appears to be the most efficient RT primer for our target and have amplified products more than 2.5 kb away from the second polyadenylation signal.

### Optimization of the PCR

The PCR must be optimized for each template to be amplified. Primer selection is the most critical part of optimization of the PCR. We choose primers that are 20–30 bases long with a 50–60% GC content. The 3' end of each primer should have 2–3 Gs and/or Cs to increase the specificity of the binding. If information about the 3' UTR region of the cDNA is available, it is helpful to place the antisense primer in this region as this confers the most specificity. As the sample contains genomic DNA, it is helpful if the genomic sequence is available. If possible, it is best to locate primers in separate but contiguous exons. Thus, if amplification of genomic DNA occurs, the band would have a higher molecular weight due to the presence of the intron sequence. The maximum efficiency of the PCR reaction is related to length of the product, therefore we generally try to position primers 200–700 bases apart. We usually choose annealing temperatures 55–60 °C. The calculated annealing temperatures of the primer pairs should be within 5 °C of each other.

Primer selection has been one of the major obstacles for the amplification of PDG. Our original primer pairs were PDG 1–2 (fig. 1). When this primer pair was used, we intermittently would get a band of the appropriate size. We suspected that the potential stem-loop structure 5' to the antisense primer (primer 1) interfered with RT. We tried various techniques to reduce secondary structures by denaturing the RNA with variable success. Next, we tried primer No. 5 as the antisense primer and finally achieved consistent strong amplification. Unfortunately, the genomic sequence for glutaminase was not known. When we attempted amplification in the absence of RT, a band of the same size was obtained, albeit much fainter than with RT. Finding an appropriate primer pair has consisted of marching up the 5' end of the molecule until we no longer obtained a band without RT. We now have a primer pair which yields a 322-bp product with RT and no bands without RT (primers 9 and 10, fig. 1).

Besides optimization of PCR primers, there are several other parameters that need to be adjusted including annealing temperature, magnesium concentration, and cycle number. The annealing temperature we have found to be optimum for PDG is 60 °C. The other parameters are still being assessed. In addition, an assessment of the efficiency, sensitivity, and precision of the PCR needs to be performed.

## Methods of Detection

Several standard methods of detection and quantification of the PCR products have been used. All methods require electrophoresis to size fractionate the products formed. The most commonly used method is a simple agarose for polyacrylamide gel electrophoresis. There are currently three methods of quantitating products: densitometry of ethidium bromide-stained gels; densitometry of autoradiograms made from either the gel directly via use of radiolabelled primers or deoxynucleotide triphosphates (dNTP) or from a Southern blot, and direct liquid scintillation counting of the product which has been labelled either with a radiolabelled dNTP or primer. Fluorescent labels have also been used although not in our laboratory. A newer method of detection and quantification is capillary electrophoresis. We are currently developing the use of capillary electrophoresis to analyze and quantitate PCR products (fig. 2). The detector measures absorbance at 260 nm and the quantity of the PCR product can be obtained from the area under the peak. Identifying a band of the expected size, while reassuring, is insufficient evidence that the product is the desired one. Confirmation of the identity of the product can be made by Southern hybridization of the gel and subsequent probing with a labelled oligo specific for the target, restriction enzyme analysis or direct sequencing of the product.

It has become popular to carry out absolute quantification of PCR products. There are however several problems with these techniques as they are currently employed. Most have utilized a mutant cDNA as an internal primer. This mutant may have an altered restriction enzyme site or an insertion or de-

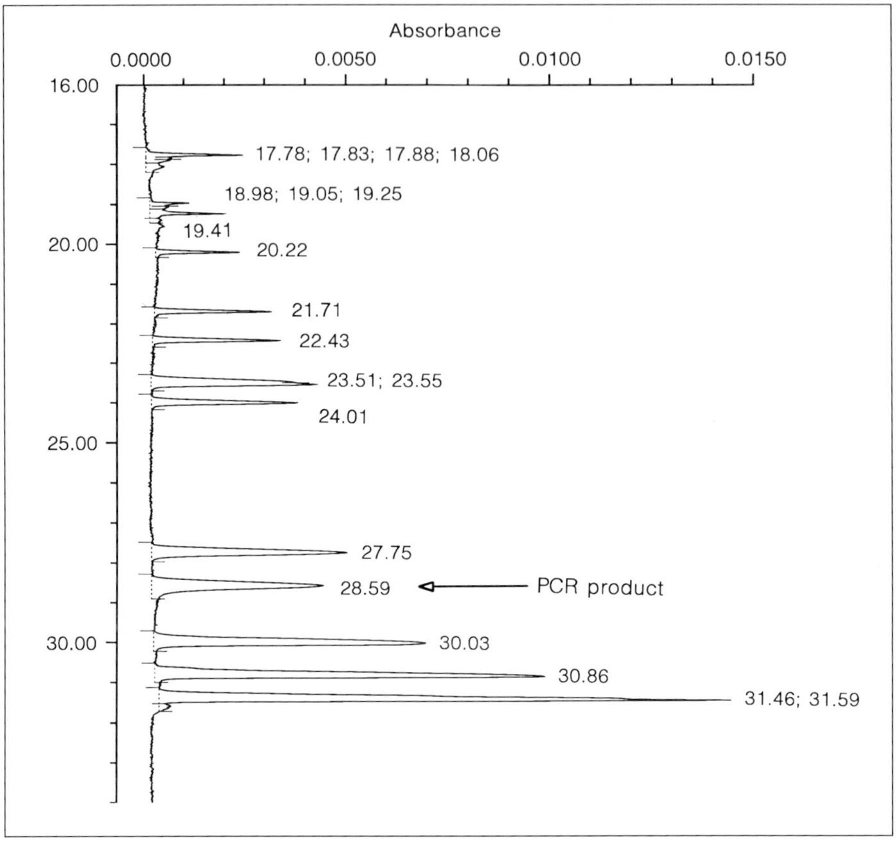


*Fig. 2.* Typical capillary electrophoresis electropherogram of a sample of a PCR mixed with a HaeIII digest of ΦX 174RF. The PCR product is a 670-bp product for aldose reductase.

letion to facilitate its distinction from the target product. Restriction sites have the problem that PCR errors leading to alteration of the restriction site would lead to an overestimation of target DNA. Insertion and deletions do not have this problem, however, all three methods do not account for efficiency of the RT step. Methods employing an internal cRNA have the advantage of also controlling for the RT step but also are more difficult to prepare and because cRNA is much less stable than cDNA, verification that the cRNA is intact must be done prior to each experiment. In addition, we believe that absolute quantification is unnecessary for most applications. The physiologic significance of the number of copies of mRNA in a given cell remains unclear given our present state of knowledge. Therefore, a relative quantification to compare the amount of message from one segment to the next or in the present case, control versus acid-loaded animals is sufficient providing that the amplification remains in the exponential phase of the PCR.

## References


1 Schoolwerth AC: Regulation of renal ammoniagenesis in metabolic acidosis. Kidney Int 1991;40:961–973.

2 Curthoys NP, Lowry OH: The distribution of glutaminase isoenzymes in the various structures of the nephron in normal, acidotic, and alkalotic rat kidney. J Biol Chem 1973;248:162–168.

3 Tong J, Harrison G, Curthoys NP: The effect of metabolic acidosis on the synthesis and turnover of rat renal phosphate-dependent glutaminase. Biochem J 1986;233:139–144.

4 Wright PA, Knepper MA: Phosphate-dependent glutaminase activity in rat renal cortical and medullary tubule segments. Am J Physiol 1990;259:F961–F970.

5 Schoolwerth AC, Nazar BL, LaNoue K: Glutamate dehydrogenase activation and ammonia formation by rat kidney mitochondria. J Biol Chem 1978;253:6177–6183.

6 Wright PA, Knepper MA: Glutamate dehydrogenase activities in microdissected rat nephron segments: Effects of acid-base loading. Am J Physiol 1990;259:F53–F59.

7 Hwang J-J, Curthoys NP: Effect of acute alterations in acid-base balance on rat renal glutaminase and phosphoenolpyruvate carboxykinase gene expression. J Biol Chem 1991;266: 9392–9396.

8 Kaiser S, Curthoys NP: Effect of pH and bicarbonate on phosphoenolpyruvate carboxykinase and glutaminase mRNA levels in cultured renal epithelial cells. J Biol Chem 1991;266: 9397–9402.

9 Kaiser S, Hwang J-J, Smith H, Banner C, Welbourne TC, Curthoys NP: Effect of altered acid-base balance and of various agonists on levels of renal glutamate dehydrogenase mRNA. Am J Physiol 1992;262:F507–F512.

10 Wright PA, Packer RK, García-Pérez A, Knepper MA: Time course of renal glutamate dehydrogenase induction during $NH_4Cl$ loading in rats. Am J Physiol 1992;262:F999–F1006.

11 Good DW, Burg MB: Ammonia production by individual segments of the rat nephron. J Clin Invest 1984;73:602–610.

12 Good DW, DuBose TD Jr: Ammonia transport by early and late proximal convoluted tubule of the rat. J Clin Invest 1987;79:684–691.

13 Shapiro RA, Farrell L, Srinivasan M, Curthoys NP: Isolation, characterization, and in vitro expression of a cDNA that encodes the kidney isoenzyme of the mitochondrial glutaminase. J Biol Chem 1991;266:18792–18796.

14 Briggs JP, Todd-Turla K, Schnermann JB, Killen PD: Approach to the molecular basis of nephron heterogeneity: Application of reverse transcription-polymerase chain reaction to dissected tubule segments. Semin Nephrol 1993;13:2–12.

15 Moriyama T, Murphy HR, Martin BM, García-Pérez A: Detection of specific mRNAs in single nephron segments by use of the polymerase chain reaction. Am J Physiol 1990;258:F1470–1474.



Susan R. DiGiovanni, MD, National Institutes of Health,
LKEM/NHLBI, Bldg. 10, Rm 6N307, Bethesda, MD 20892 (USA)

Tizianello A, Baverel G, Endou H, Schoolwerth AC, O'Donovan DJ (eds):
Renal Ammoniagenesis and Interorgan Cooperation in Acid-Base Homeostasis.
Contrib Nephrol. Basel, Karger, 1994, vol 110, pp 88–97

# Ammoniagenesis in Renal Cell Culture

## A Comparative Study on Ammonia Metabolism of Renal Epithelial Cell Lines[1]

*Gerhard J. A. Gstraunthaler*[2]

Institute of Physiology, University of Innsbruck, Austria

The initial reactions in the primary pathway of renal metabolism of glutamine and renal ammoniagenesis, respectively, are catalyzed by phosphate-dependent glutaminase (PDG) and glutamate dehydrogenase (GDH) [1] (fig. 1). In addition, transamination pathways have been described [2, 3], where alanine or aspartate can be formed from glutamate, either by alanine aminotransferase (ALT) or aspartate aminotransferase (AST).

In recent years, renal epithelial cell cultures [4] have been successfully introduced as valuable model systems to study renal ammoniagenesis and its acid-base regulation in vitro [5]. In this context, the LLC-$PK_1$ pig renal epithelial cell line was most widely used [6–11]. We recently characterized the ammoniagenic pathways in LLC-$PK_1$ cells and their adaptive response to metabolic acidosis [11]. LLC-$PK_1$ cultures readily adapted to metabolic acidosis with an increase in glutamine consumption and ammonia production. In addition, alanine was found as an end product of LLC-$PK_1$ glutamine metabolism, indicating that transamination constitutes a major pathway in LLC-$PK_1$ ammoniagenesis [11]. Thus, the LLC-$PK_1$ ammoniagenic pathways resemble the situation in dog kidney, where in contrast to the rat, the metabolic flux from glutamate to α-ketoglutarate is performed via transamination by ALT, rather than deamination by GDH (fig. 1).

[1] This study was supported in part by the Dr. Legerlotz Foundation, and by the Austrian Science Foundation, grant No. P8074.

[2] With the technical assistance of Gertraud Vetter.

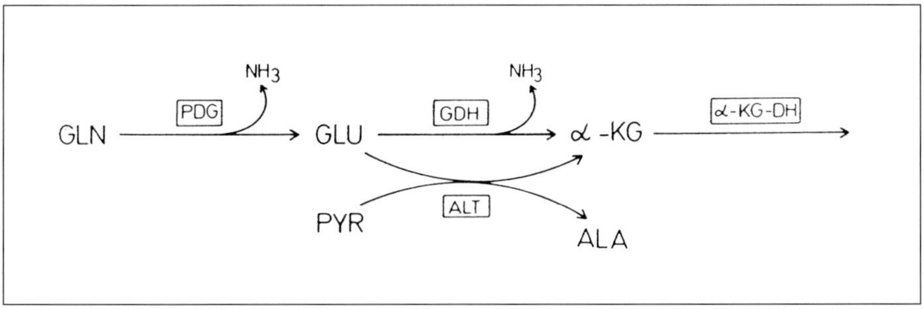


*Fig. 1.* Primary pathways of proximal tubular glutamine metabolism and renal ammoniagenesis. PDG = Phosphate-dependent glutaminase; GDH = glutamate dehydrogenase; α-KG-DH = oxoglutarate dehydrogenase; ALT = alanine aminotransferase.

In order to delineate possible species differences in glutamine metabolism retained by renal epithelial cells in tissue culture, in the present study, LLC-$PK_1$ cultures were compared with other proximal tubular-like renal epithelial cell lines, the OK (opossum kidney) line [4], and NRK (normal rat kidney) cells, the latter of which have already been used in in vitro ammoniagenesis studies [12, 13]. In addition, MDCK (Madin-Darby canine kidney) cells were investigated, although these cells have been described to be of collecting duct origin [14, 15].

## Materials and Methods

*Cell Culture*

LLC-$PK_1$ (ATCC CRL 1392), NRK-52E (ATCC CRL 1571) and MDCK (ATCC CCL 34) cells were purchased from American Type Culture Collection, Rockville, Md., USA. OK cells were kindly provided by Dr. David Warnock, San Francisco, Calif., USA. All cell lines were maintained in Dulbecco's modified Eagle's medium (DMEM) with 5.5 m*M* *D*-glucose and 2 m*M* *L*-glutamine, supplemented with 10% fetal bovine serum, 100 U/ml penicillin, and 100 μg/ml streptomycin as described elsewhere [11, 14–19].

Metabolic acidosis in tissue culture (pH 7.0) was generated by lowering the bicarbonate concentration in culture medium to 9.0 m*M* at constant $CO_2$ atmosphere, as described in detail [11, 19]. All experiments were performed with 10- to 14-day confluent monolayer cultures.

*Assays*

Enzyme activities were determined in whole cell homogenates as described [11, 16–18]. Metabolites in culture media (glucose, lactate, glutamine, ammonia, alanine, aspartate) were analyzed enzymatically by established methods [11, 16–19].

*Table 1.* Enzyme activities of key ammoniagenic enzymes expressed in renal epithelial cell lines

| Enzyme | Cell line | | | |
|---|---|---|---|---|
| | LLC-$PK_1$ | OK | NRK | MDCK |
| PDG | 23.4 ± 2.6 (15) | 42.5 ± 2.5 (6) | 41.5 ± 11.9 (6) | 34.3 ± 3.4 (4) |
| GDH | 463 ± 53 (12) | 681 ± 76 (7) | 438 ± 77 (6) | 266 ± 34 (5) |
| ALT | 112 ± 7.2 (5) | 26.8 ± 3.5 (5) | 7.4 ± 1.1 (7) | 4.8 ± 0.8 (5) |
| AST | 404 ± 57 (6) | 1,262 ± 113 (5) | 593 ± 74 (6) | 257 ± 18 (6) |

Enzyme activities are expressed as mU/mg protein (nmol/min/mg cell protein). Values are given as mean ± SD (n). PDG = Phosphate-dependent glutaminase; GDH = glutamate dehydrogenase; ALT = alanine aminotransferase; AST = aspartate aminotransferase.

*Table 2.* Ammonia ($NH_3$) and *L*-alanine (ALA) baseline production rates by renal epithelial cell lines

| Metabolite | Cell line | | | |
|---|---|---|---|---|
| | LLC-$PK_1$ | OK | NRK | MDCK |
| $NH_3$ | 1.09 ± 0.13 (11) | 1.93 ± 0.06 (8) | 1.71 ± 0.16 (6) | 0.62 ± 0.06 (6) |
| ALA | 0.96 ± 0.05 (6) | 0.35 ± 0.09 (6) | 1.18 ± 0.16 (6) | 0.12 ± 0.10 (4) |
| Ratio $NH_3$:ALA | 1:0.88 | 1:0.18 | 1:0.69 | 1:0.19 |

Metabolite production rates are expressed as µmol/mg protein and 24 h at pH 7.6. Values are given as mean ± SD (n).

## Results

### *Basic Characterization of Cell Lines*

In table 1, the activities of the key ammoniagenic enzymes PDG, GDH, ALT, and AST, expressed in renal epithelial cell cultures under control conditions (pH 7.6) are summarized. OK cells expressed the highest activities of PDG, GDH, and AST, whereas ALT activity was considerably lower compared with LLC-$PK_1$ cells. NRK cells, due to their rat origin, where in rat kidney in vivo virtually no ALT activity is expressed [20, 21], also exhibited very low ALT levels. AST activity was expressed in all cell lines tested, with OK cells expressing by far the highest activity values.

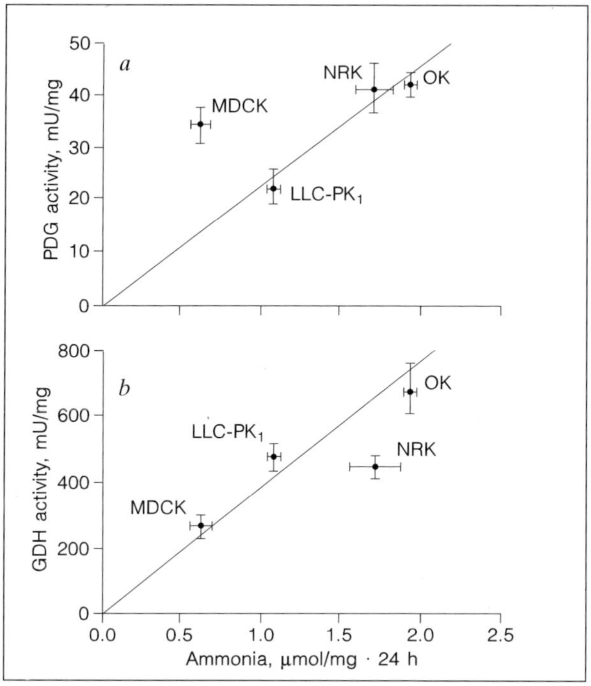


*Fig. 2.* Correlation of baseline ammonia production rates of the four renal epithelial cell lines with expression of PDG *(a)* and GDH activities *(b)*. Data were taken from tables 1 and 2, respectively.

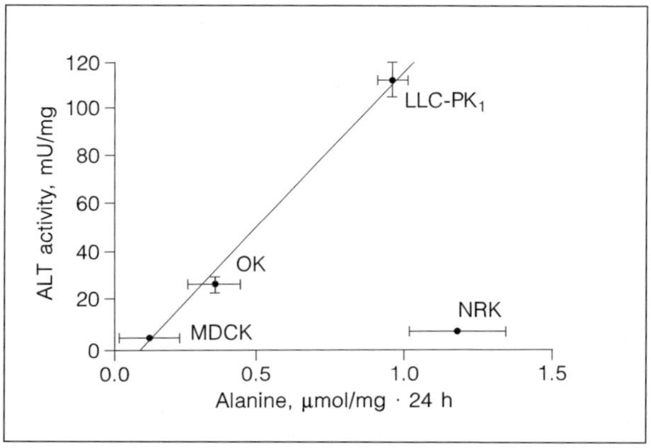


*Fig. 3.* Correlation of basal rates of alanine production with expression of ALT activities. For details see figure 2.

In parallel, baseline production rates of ammonia and *L*-alanine of the renal cell lines were determined in 24-hour intervals from 2 m*M* *L*-glutamine at pH 7.6. As depicted in table 2, OK cells excreted the highest amounts of ammonia into the tissue culture medium, paralleled by very low alanine levels, which is in congruence with the elevated PDG and GDH activites and the rather low ALT activity found in OK cells (table 1). NRK cells also generated high

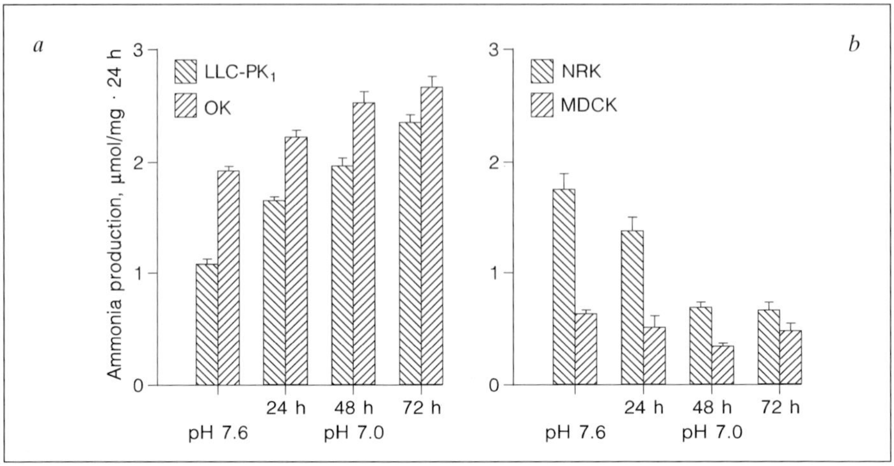


*Fig. 4a, b.* Ammonia production by renal epithelial cell cultures in response to metabolic acidosis (24–72 h at pH 7.0). Metabolic rates are expressed as μmol/mg cell protein × 24 h. Values are means ± SD (n = 6–12). For control values (pH 7.6) see also table 2.

amounts of ammonia, and, in contrast to the low ALT activity, also a high rate of alanine production was found. Thus, in NRK cells alanine must been formed in pathways other than transamination by ALT. LLC-$PK_1$ cultures produced ammonia and alanine at a molar ratio of 1:1 of 1 μmol/mg·24 h, as already described in detail [11]. In all four cell lines investigated, no aspartate excretion into culture media could be detected, besides fairly high AST activities (table 1).

In figures 2 and 3, the baseline production rates of ammonia and alanine by the four cell lines were correlated with the expression of the corresponding enzyme activities PDG, GDH, and ALT, respectively. As shown in figure 2, basal ammonia production rates correspond fairly well with the enzyme acitivities of PDG as well as GDH. Comparing basal alanine production rates with the expression of ALT activities (fig. 3) revealed straight-line correlations for LLC-$PK_1$, OK and MDCK cells, with LLC-$PK_1$ cells exhibiting the highest rates of alanine production and ALT activity expression, respectively. The high rate of alanine production by NRK cells despite extremely low ALT activity, as described above, can also be well recognized in figure 3.

*Metabolic Acidosis*

Metabolic acidosis in confluent monolayer cultures was elicited by lowering the bicarbonate content in tissue culture medium at a constant $CO_2$ content in the incubator atmosphere [11], which resulted in a decrease of culture medium pH from 7.6 to 7.0. The metabolic response in ammonia production of

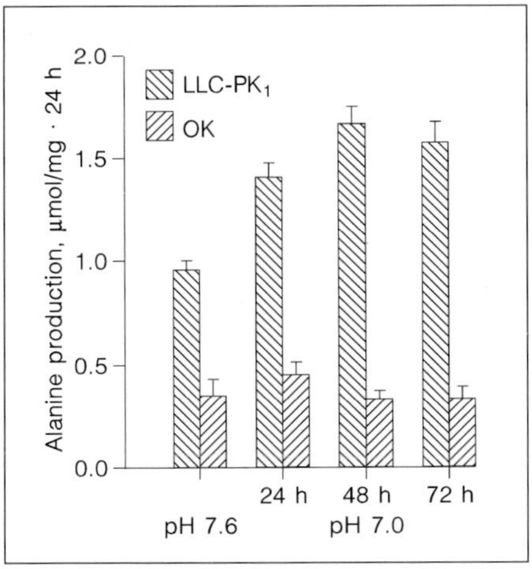


*Fig. 5.* Rates of alanine production by LLC-$PK_1$ and OK cultures under control and acidotic conditions. For details see figure 4.

LLC-$PK_1$, OK, NRK and MDCK cells upon acidotic culture conditions for up to 72 h is presented in figure 4. Confluent cultures were switched to acidic culture conditions, and culture media were changed every 24 h. As can be seen for LLC-$PK_1$ and OK cells (fig. 4a), after 24–72 h of adaptation the increase in ammonia formation in dependence of culture medium pH is well pronounced. OK cells start with higher basal ammonia production rates, compared to LLC-$PK_1$ cells (see also table 2), which gradually increase during the 72-hour adaptation period. The percentage increase in ammonia excretion, however, is higher in LLC-$PK_1$ cells, resulting in a doubling of the rate of ammonia generation in comparison to controls (pH 7.6) [11]. On the contrary, NRK and MDCK cultures did not show any adaptive response in ammonia metabolism upon metabolic acidosis. As depicted in figure 4b, 24-hour ammonia production rates of NRK cells even decreased, or remained unchanged in MDCK cultures, during a 72-hour acidosis. Adaptive increases in alanine production rates were only seen in LLC-$PK_1$ cultures [11], whereas in OK cells alanine generation remained unaltered at low baseline values (fig. 5, table 2).

In order to monitor the time course of ammonia and alanine accumulation in tissue culture medium, confluent LLC-$PK_1$ and OK cultures were incubated at pH 7.6 and pH 7.0, and metabolites determined every 24 h in culture media samples. The results are summarized in figure 6. During the first 24 h, the time course of ammonia generation under control as well as acidotic conditions is almost identical in both cell lines, with higher rates at pH 7.0 compared with pH 7.6. Thereafter, ammonia production by LLC-$PK_1$ cells continued, whereas in OK cultures a plateau phase was reached (fig. 6a). Alanine accumulated in media of LLC-$PK_1$ cultures with ammonia at equimolar amounts in

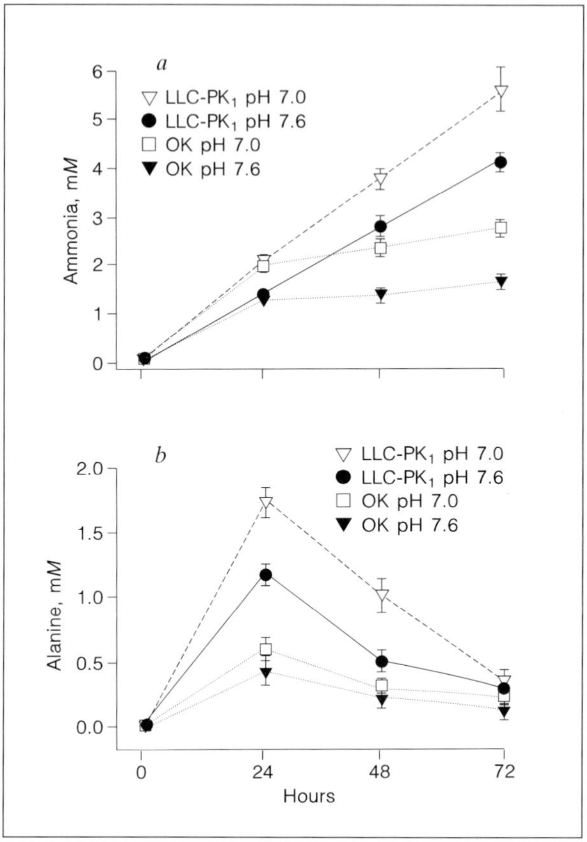


*Fig. 6.* Time course of ammonia *(a)* and alanine production *(b)* of LLC-$PK_1$ and OK cultures under control conditions (pH 7.6) and in metabolic acidosis (pH 7.0). Confluent monolayer cultures were incubated for 72 h as indicated. At 24-hour intervals, culture medium samples (100 µl) were taken and analyzed.

the first 24 h (see also fig. 5); however, in the next 48 h of continuous incubation, alanine was completely taken up by the cells (fig. 6b). Similar time courses, although at lower levels, were found for OK cells, with almost no differences between controls and acidotic cultures.

PDG, GDH, and ALT activities did not undergo any adaptive changes in OK cells during the 24- to 72-hour adaptation period, which is identical with the situation in LLC-$PK_1$ cells [11].

## Discussion

In the present study, ammonia metabolism and its regulation by metabolic acidosis was investigated in the LLC-PK1, OK, NRK and MDCK renal epithelial cell lines [4] in order to delineate species differences in the ammoniagenic pathway retained by the cells in tissue culture, and to test for the suitability and usefulness of the cell lines as possible in vitro model systems in studying renal ammoniagenesis at the cellular and subcellular level.

All four cell lines express the key ammoniagenic enzymes PDG, GDH and ALT (table 1), whose enzyme activities correspond fairly well with the baseline production rates of ammonia and alanine (table 2, fig. 2, 3). The only exception is found in NRK cells, which exhibited high alanine production rates, despite extremely low ALT activity level, which might be due to their rat renal origin, since in rat kidney in vivo virtually no ALT activity is expressed [20, 21]. Therefore, alanine must be generated in NRK cells in pathways other than transamination by ALT. Recently, a possible role of the glutaminase II pathway [1, 11], catalyzed by glutamine ketoacid aminotransferase and ω-amidase, was discussed for alanine generation in NRK cells [12].

Upon metabolic acidosis in tissue culture, only LLC-$PK_1$ and OK cells responded with an increase in ammonia production. OK cells express higher PDG and GDH activities under control conditions, accompanied by higher baseline ammonia production rates, than do LLC-$PK_1$ cells (tables 1, 2). Together with the observed low ALT activity in OK cells, these findings can be interpreted as a shift towards enhanced deamination by GDH in OK cell glutamine metabolism (fig. 1), which is further supported by the very low basal alanine production (see below). However, OK cells did not that readily adapt to pH 7.0 by increased ammonia formation as did LLC-$PK_1$ cells (fig. 4a). LLC-$PK_1$ cells have already been described as valuable cell culture model systems to study renal ammoniagenesis in vitro [6–11]. We [11] and others [6–10] showed that LLC-$PK_1$ cells respond to acidotic culture conditions by an adaptive increase in glutamine consumption and ammonia production, accompanied by an equimolar generation of alanine, which was interpreted that transamination appears to be a main ammoniagenic pathway in LLC-$PK_1$ cells [11]. In OK cells low baseline production rates of alanine were found, which correlated well with the decreased expression of ALT activity (fig. 3), indicating a minor role for transamination in OK cell ammoniagenesis. Furthermore, alanine generation by OK cells did not show any adaptive changes upon culturing the cells at pH 7.0 for 24–72 h (fig. 5, 6). In both cell lines, no adaptive changes in PDG, GDH or ALT activities could be observed [11].

Recently, the time course of glutamine consumption and ammonia production was monitored under control conditions (pH 7.6) in LLC-$PK_1$ and OK cultures [11]. In the present study, these experiments were repeated, and extended to acidotic conditions (pH 7.0), and to the monitoring of *L*-alanine. The results revealed at pH 7.6 and 7.0 identical ammonia production rates of both cell lines for the first 24-hour interval (fig. 6a). Thereafter, ammonia generation and accumulation in tissue culture medium, respectively, continuously increased in LLC-$PK_1$ cultures, although, as shown previously [11], glutamine was completely taken up by the cells after 24–48 h. Ammonia formation of OK cells reached a plateau phase, again at different levels in dependence of the pH of the culture medium. These differences in ammonia generation by LLC-$PK_1$ and OK cells may be based on the formation, and subsequent cellular re-entry and metabolism of alanine, respectively (fig. 6b). After excretion of alanine by LLC-$PK_1$ cells into tissue culture medium in a 1:1 ratio with ammonia after 24 h [11], alanine is consumed thereafter during 48–72 h of continuous incubation, where it might serve as ammoniagenic amino acid.

Recently, NRK cells were used as model system in in vitro ammoniagenesis studies [12, 13]. In contrast to the findings in the present study, where ammonia production by NRK cells decreased during a 24- to 72-hour acidosis (fig. 4), Nissim et al. [12] described a significant increase in ammonia production from either glutamine or glutamate at pH 7.0 vs. 7.4.

## Conclusion

In summary, the proximal tubule-like renal epithelial cell lines LLC-$PK_1$ and OK may be used as valuable model systems to study renal ammoniagenesis and its acid-base regulation in vitro. From both cell lines, gluconeogenic substrains have recently been isolated [16, 18], which may provide additional tools in studies of renal proximal tubular metabolism at the cellular and subcellular level [22, 23].

MDCK cells, due to their collecting duct origin [14], do not respond with adaptive changes in ammonia metabolism upon acidosis. However, this cell line has recently been shown to differentiate in acidotic (and hyperosmotic) culture conditions into acid secreting-like intercalated cells [15, 24], and may therefore serve as powerful cell culture model for acid-base changes in distal nephron and collecting duct portions.

## References


1 Tannen RL, Sahai A: Biochemical pathways and modulators of renal ammoniagenesis. Miner Electrolyte Metab 1990;16:249–258.
2 Vinay P, Lemieux G, Gougoux A, Halperin M: Regulation of glutamine metabolism in dog kidney in vivo. Kidney Int 1986;29:68–79.
3 Brosnan JT, Lowry M, Vinay P, Gougoux A, Halperin ML: Renal ammonium production – une vue canadienne. Can J Physiol Pharmacol 1987;65:489–498.
4 Gstraunthaler GJA: Epithelial cells in tissue culture. Renal Physiol Biochem 1988;11:1–42.
5 Nissim I, Korzets Z: Ammonia metabolism in cultured renal cells. Miner Electrolyte Metab 1990;16:270–276.

6 Cole LA, Scheid JM, Tannen RL: Induction of mitochondrial metabolism and pH-modulated ammoniagenesis by rocking LLC-$PK_1$ cells. Am J Physiol 1986;251:C293–C298.
7 Sahai A, Cole LA, Tannen RL: Pathways and regulation of ammoniagenesis by the LLC-$PK_1$ cells in culture. J Lab Clin Med 1989;114:285–293.
8 Sahai A, Cole LA, Clarke DL, Tannen RL: Rocking promotes differentiated properties in LLC-$PK_1$ cells by improved oxygenation. Am J Physiol 1989;256:C1064–C1069.
9 Sahai A, Laughrey E, Tannen RL: Relationship between intracellular pH and ammonia metabolism in LLC-$PK_1$ cells. Am J Physiol 1990;258:F103–F108.
10 Sahai A, Nissim I, Tannen RL: Pathways of acute pH regulation of ammoniagenesis in LLC-$PK_1$ cells: Study with [$^{15}$N]glutamine. Am J Physiol 1991;261:F481–F487.
11 Gstraunthaler G, Landauer F, Pfaller W: Ammoniagenesis in LLC-$PK_1$ cultures: Role of transamination. Am J Physiol 1992;263:C47–C54.
12 Nissim I, States B, Yudkoff M, Segal S: Characterization of amino acid metabolism by cultured rat kidney cells: Study with $^{15}$N. Am J Physiol 1987;253:F1243–F1252.
13 Nissim I, Yudkoff M, Nissim I, States B: Effects of 5-amino-4-imidazolecarboxamide riboside on the purine nucleotide synthesis and growth of rat kidney cells in culture: Study with [$^{15}$N]aspartate. J Cell Physiol 1989;138:536–540.
14 Pfaller W, Gstraunthaler G, Kersting U, Oberleithner H: Carbonic anhydrase activity in Madin-Darby canine kidney cells. Evidence for intercalated cell properties. Renal Physiol Biochem 1989;12:328–337.
15 Pfaller W, Steigner W, Gstraunthaler G, Kersting U, Oberleithner H: Low pH and hyperosmolality determine the differentiation of MDCK cells typical for the outer medullary collecting duct. Cell Physiol Biochem 1993;3:56–67.
16 Gstraunthaler G, Handler JS: Isolation, growth, and characterization of a gluconeogenic strain of renal cells. Am J Physiol 1987;252:C232–C238.
17 Gstraunthaler G, Gersdorf E, Fischer WM, Joannidis M, Pfaller W: Morphological and biochemical changes of LLC-$PK_1$ cells during adaptation to glucose-free culture conditions. Renal Physiol Biochem 1990;13:137–153.
18 Gstraunthaler G, Thurner B, Weirich-Schwaiger H, Pfaller W: A novel gluconeogenic strain of OK cells with metabolic properties different from gluconeogenic LLC-$PK_1$ cells. Cell Physiol Biochem 1993;3:78–88.
19 Gstraunthaler G, Landauer F, Pfaller W: Ammoniagenesis in renal cell culture. Lack of extracellular ammoniagenesis at the apical surface of LLC-$PK_1$ epithelia. Renal Physiol Biochem 1993;16:203–211.
20 Szepesi B, Avery EH, Freedland RA: Role of kidney in gluconeogenesis and amino acid catabolism. Am J Physiol 1970;219:1627–1631.
21 Chan AWK, Perry SG, Burch HB, Fagioli S, Alvey TR, Lowry OH: Distribution of two aminotransferases and *D*-amino acid oxidase within the nephron of young and adult rats. J Histochem Cytochem 1979;27:751–755.
22 Kaiser S, Curthoys NP: Effect of pH and bicarbonate on phosphoenolpyruvate carboxykinase and glutaminase mRNA levels in cultured renal epithelial cells. J Biol Chem 1991;266: 9397–9402.
23 Kaiser S, Hwang J-J, Smith H, Banner C, Welbourne TC, Curthoys NP: Effect of altered acid-base balance and of various agonists on levels of renal glutamate dehydrogenase mRNA. Am J Physiol 1992;262:F507–F512.
24 Kersting U, Schwab A, Treidtel M, Pfaller W, Gstraunthaler G, Steigner W, Oberleithner H: Differentiation of Madin-Darby canine kidney cells depends on cell culture conditions. Cell Physiol Biochem 1993;3:42–55.

Gerhard J. A. Gstraunthaler, PhD, Institute of Physiology, University of Innsbruck,
Fritz-Pregl-Straße 3, A-6010 Innsbruck (Austria)

Tizianello A, Baverel G, Endou H, Schoolwerth AC, O'Donovan DJ (eds):
Renal Ammoniagenesis and Interorgan Cooperation in Acid-Base Homeostasis.
Contrib Nephrol. Basel, Karger, 1994, vol 110, pp 98–102

# Ammonia Production in Cell Lines Established from Transgenic Mice Harboring Temperature-Sensitive Simian Virus 40 Large T-Antigen Gene

*Takashi Sekine*[a], *Makoto Hosoyamada*[a], *Akiko Haga-Mizuno*[a], *Michio Takeda*[a], *Misao Suzuki*[b], *Masao Obinata*[c], *Hitoshi Endou*[a]

[a] Department of Pharmacology and Toxicology, Kyorin University, Tokyo;
[b] Kumamoto University, Kumamoto;
[c] Tohoku University, Sendai, Japan

Since the nephron consists of a variety of extremely heterogeneous cells, culture cells may provide new approaches for physiological and biochemical studies of the nephron. Although several renal cell lines, such as LLC-$PK_1$ and MDCK, have been used [1], their precise origins remain hypothetical. An alternative to established cell lines is the use of primary culture cells from a well-defined nephron segment. However, studies with primary culture cells are also limited by changes of their wild characters according to further passage, and their limitation of lifespan makes it impossible to obtain sufficient numbers of cells. Establishment of immortalized cell lines of well-defined nephron segments retaining the differentiated properties is ideal in this respect.

In the present study, we report the established cell lines from individual nephron segments of transgenic mice harboring temperature-sensitive simian virus 40 (ts-SV40) large T-antigen gene. Using these cell lines, ammoniagenesis was investigated as one of the differentiated renal cell functions.

## Materials and Methods

*Establishment of Renal Cell Lines from Transgenic Mice.* A ts-SV40 large T-antigen sequence was derived from pSVtsA58(ori$^-$). *Bam*HI fragments of SV40tsA58 mutant gene that had been subcloned in pBR322 were injected into pronuclei of fertilized eggs of C57BL/6 mice, and the eggs were fostered in pseudopregnant ICR mice. The presence of intact SV40 large T-antigen genes was analyzed by Southern blot hybridization in each newborn mouse [2].

To obtain renal cell lines, nephron segments including the first segment of the proximal tubule (S1) and the outer medullary collecting tubule (OMCT) were microdissected from the transgenic mice by previously described methods [3]. Individual nephron segments microdissected under a sterile condition were cultured on collagen-coated plastic dishes. The culture medium, RITC 80-7 (Kyokuto Pharmaceutical Industrial Co. Ltd, Tokyo, Japan) was supplemented with 5% fetal bovine serum, 10 μg/ml transferrin, 1 μg/ml insulin and 10 ng/ml recombinant epidermal growth factor [2, 4]. These cells grew at a permissive temperature (33 °C), but not at a nonpermissive temperature (39 °C).

*Determination of Ammonia.* The methods of determination of ammonia produced by freshly dissected mice nephron segments from *L*-glutamine were as described previously [3]. The ammonia produced by culture cells from *L*-glutamine was quantified by an enzymatic method [5]. Briefly, the culture medium with and without 2 m*M* *L*-glutamine were deproteinized with an equal volume of 10% trichloroacetic acid, and then neutralized by 2 *M* $KHCO_3$ after 1, 6, 12 and 24 h of incubation. The ammonia in the culture medium was then converted to NAD in the presence of 2-oxoglutarate, NADH, and glutamate dehydrogenase. The decrease of NADH, as measured by the change of extinction at 340 nm, was estimated as the amount of ammonia. The ammoniagenic activity was calculated as difference between with and without *L*-glutamine.

*Measurement of Protein Content.* Protein content of each cell line was measured by the method of Lowry et al. [5].

*Effect of CDDP on Ammoniagenesis.* The effect of cisplatin (CDDP) on ammoniagenesis was assessed by a culture cell line derived from the proximal S1 segment. CDDP was added to the cells in 24-well dishes (final concentration $5 \times 10^{-7}$ to $5 \times 10^{-4}$ *M*). Ammonia produced from 2 m*M* *L*-glutamine in the culture medium was measured after a further 24-hour culture.

## Results and Discussion

In this study, we examined ammoniagenesis as one of the differentiated characters in renal cells. Figure 1 shows the time course of ammonia production by S1 and OMCT culture cell lines. After 6, 12 and 24 h, the amounts of ammonia produced by the S1 cell line were 6.22, 8.22 and 11.63 nmol/μg tissue protein respectively, and were about 2–3 times higher than those of the OMCT cell line. Figure 2 shows the ammonia production in freshly dissected mice nephron segments. The amount of ammonia produced by the freshly isolated S1 segment was calculated as 5.82 nmol/μg protein/6 h, which is almost equivalent to that of the S1 culture cell for the same hours (6.22 nmol/μg). These data show that ammoniagenic activity is preserved in the present cell lines. These cell lines may provide the possibility to investigate various site-specific differentiated kidney functions, including the intracellular regulatory mechanism of renal ammoniagenesis.

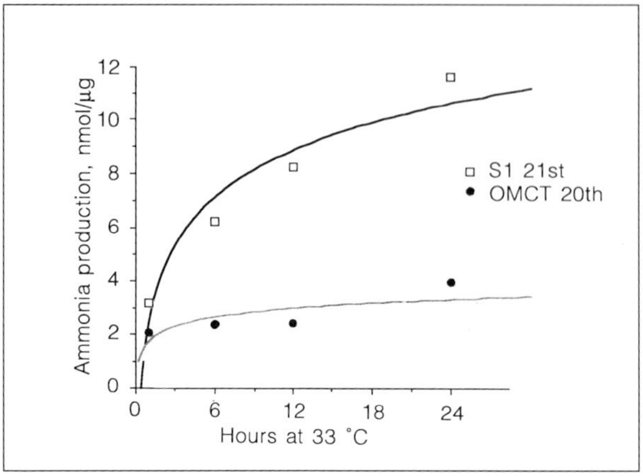


*Fig. 1.* Ammoniagenesis from 2 m*M* *L*-glutamine in immortalized mouse cells. Each point represents mean value of triplicate determinations.

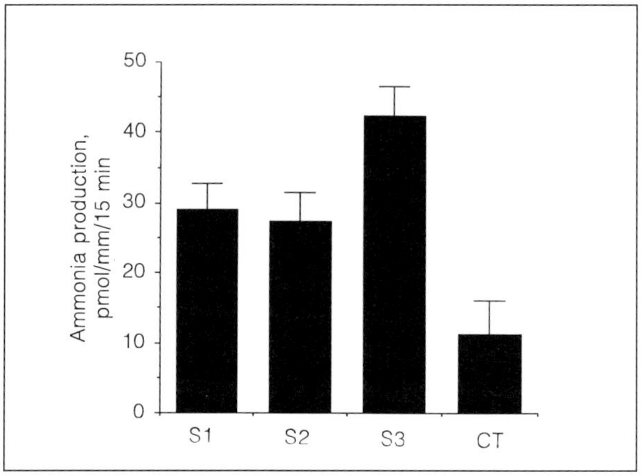


*Fig. 2.* Ammonia production in nephron segments microdissected from mouse kidneys. Columns and respective vertical bars show means ± SE.

Concerning establishment of cell lines, some types of viral oncogenes are known to display immortalizing activity and the introduction of one of these genes to a primary culture cell can yield an immortalized cell line [2]. SV40 large T-antigen is one such gene that has been most widely used. Several studies have suggested that immortalization of various types of cells by introducing SV40 gene to primary culture cells could lead to some stabilization of cell-

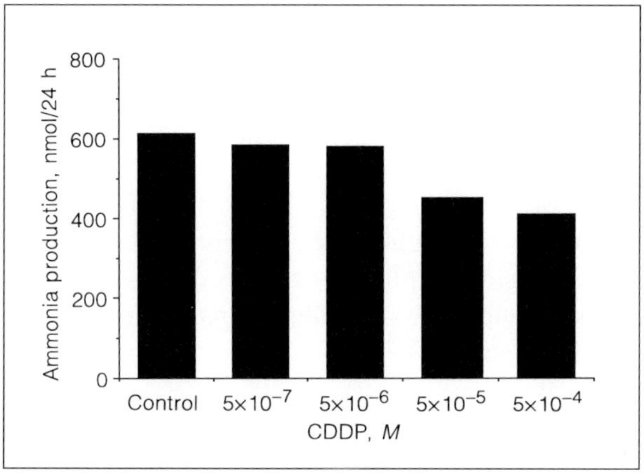


*Fig. 3.* Effect of CDDP on ammoniagenesis in proximal S1 cell line. Values are means of triplicate determinations.

specific functions [6–9]. Recently, using transgenic technology, immortalized cell lines were also obtained from transgenic mice for SV40 large T-antigen gene [10, 11].

In the present study, we used transgenic mice harboring ts-SV40 large T-antigen, and established renal cell lines from primary culture cell lines of individual nephron segments. Because the promotor used is that of SV40 large T-antigen gene itself, this gene is thought to be ubiquitously expressed in the transgenic mice. So various cell lines can be obtained from the same mouse at the same time. At the present time the cell lines established were as follows: glomerulus, proximal S1, S2 and S3 segments, medullary thick ascending limb of Henle, cortical thick ascending limb of Henle, distal tubule (DT), cortical collecting tubule (CCT), OMCT and inner medullary collecting duct. Hepatic cell lines have already been established from the same type of transgenic mice and shown liver-specific morphological and biochemical properties [2].

CDDP is known to possess serious nephrotoxicity. Clinically, CDDP-induced acute renal failure has been associated with enzymuria involving N-acetyl-β-*D*-glucosaminidase and γ-glutamyltranspeptidase, urinary excretion of $\beta_2$-microglobulin and aminoaciduria, suggesting widespread proximal tubular dysfunction [12]. The effect of CDDP on ammoniagenesis is shown in figure 3. Dose-related suppression of ammoniagenesis in S1-derived culture cell was observed. In separate experiments, we assessed the difference in sensitivity to CDDP between S3 and OMCT cell lines by estimating protein content reduction of the cell, and higher susceptibility of CDDP to S3 than

OMCT cell line was observed [unpubl. observation]. Thus, these findings may also imply usefulness of our newly established cell lines in pharmacology and toxicology.

In other studies, responsiveness of $10^{-10}$ to $10^{-6}$ *M* arginine vasopressin to cell lines derived from DT, CCT and OMCT was observed, but no response was found to cells from S1 (data not shown).

These results suggest that our established cell lines retain some differentiated properties present in their original nephron segments. Though further studies are needed, these cell lines could serve as a valuable model for the studies of cellular functions of the nephron.

## References


1 Handler JS, Kreisgerg JI: Biology of renal cells in culture; in Brenner BM, Rector FC (eds): The Kidney, ed 4. Philadelphia, Saunders, 1991, vol 1, pp 110–131.

2 Yanai N, Suzuki M, Obinata M: Hepatocyte cell lines established from transgenic mice harboring temperature-sensitive simian virus 40 large T-antigen gene. Exp Cell Res 1991;197:50–56.

3 Nonoguchi H, Uchida S, Shiigai T, Endou H: Effect of chronic metabolic acidosis on ammonia production from *L*-glutamine in microdissected rat nephron segments. Pflügers Arch 1985; 403:229–235.

4 Kun E, Kearney EB: Ammonia; in Bergmeyer HU (ed): Methods of Enzymatic Analysis, ed 2. Weinheim, Verlag Chemie, 1974, vol 4, pp 1802–1806.

5 Lowry OH, Rosebrough NJ, Farr AL, Randall RJ: Protein measurement with the Folin phenol reagent. J Biol Chem 1951;193:265–275.

6 Scott DM, Zierold K, Kinne R: Development of differentiated characteristics in cultured kidney (thick ascending loop of Henle) cells. Exp Cell Res 1986;162:521–529.

7 Scott DM, Macdonald C, Brezeski H, Kinne R: Maintenance of expression of differentiated function of kidney cells following transformation by SV40 early region DNA. Exp Cell Res 1986;166:391–398.

8 Nemunaitis J, Andrews DF, Crittenden C, Kaushansky K, Singer JW: Response of simian virus 40 (SV40)-transformed, cultured human marrow stromal cells to hematopoietic growth factors. J Clin Invest 1989;83:593–601.

9 Isom HC, Tevethia MJ, Taylor JM: Transformation of isolated rat hepatocytes with simian virus 40. J Cell Biol 1980, 85:651–659.

10 Stoos BA, Náray-Fejes-Tóth A, Carretero OA, Ito S, Fejes-Tóth G: Characterization of a mouse cortical collecting duct cell line. Kidney Int 1991;39:1168–1175.

11 Paul D, Höhne M, Pinkert C, Piasecki A, Ummelmann E, Brinster RL: Immortalized differentiated hepatocyte lines derived from transgenic mice harboring SV40 T-antigen. Exp Cell Res 1988;175:354–362.

12 Nosaka K, Nakada J, Endou H: Cisplatin-induced alterations in renal structure, ammoniagenesis and gluconeogenesis of rats. Kidney Int 1992;41:73–79.



Hitoshi Endou, MD, PhD, Department of Pharmacology and Toxicology,
Kyorin University School of Medicine, 6-20-2 Shinkawa, Mitaka-shi, Tokyo 181 (Japan)

Tizianello A, Baverel G, Endou H, Schoolwerth AC, O'Donovan DJ (eds):
Renal Ammoniagenesis and Interorgan Cooperation in Acid-Base Homeostasis.
Contrib Nephrol. Basel, Karger, 1994, vol 110, pp 103–108

# Regulation of $Na^+$-Dependent Amino Acid Transport in the Bovine Renal Epithelial Cell Line NBL-1

*J. D. McGivan, S. Plakidou-Dymock*

Department of Biochemistry, University of Bristol, UK

Amino acid transport into renal brush border vesicles from various species has been extensively investigated. We have shown that a wide range of neutral amino acids including glutamate, alanine, phenylalanine and leucine are transported via a single broad-specificity $Na^+$-dependent transport system in bovine renal vesicles [1]. This broad-specificity system occurs only in renal [1] and intestinal [see 2] epithelial membranes. Glutamate is transported into bovine brush border membrane vesicles via a high-affinity $Na^+$-dependent system which is similar to the System $X_{AG}$-characterized in other cell types [3].

The bovine renal epithelial cell line NBL-1 appears to be a useful model system to study regulatory mechanisms of amino acid transport. These cells have low levels of the proximal tubule marker enzyme alkaline phosphatase and do not express $Na^+$-dependent glucose transport. However, the specificity and kinetic properties of both amino acid and phosphate transport in these cells has been shown to be very similar to those in renal brush border membrane vesicles.

## Methods

NBL-1 cells were obtained from Flow Laboratories (McLean, Va., USA) and cultured in Ham's F-12 medium containing 10% newborn calf serum [4]. Amino acid-free medium contained the inorganic salts of this medium plus 10 m*M* glucose plus 0.1% bovine albumin [5]. Amino acid transport was measured as described previously [4–6], in the presence of 0.5 m*M* amino-oxyacetate to prevent metabolism of the labelled substrate. The $Na^+$-independent rate of transport was measured in a medium where NaCl was replaced by an equal concentration of choline Cl; this was subtracted from the rate in standard medium to give the $Na^+$-dependent component of transport.

*Table 1.* Effect of amino acid deprivation and hyperosmotic medium on alanine transport (0.1 m*M*) in NBL-1 cells

| Incubation conditions | Initial rate of $Na^+$-dependent alanine uptake % of control | |
|---|---|---|
| | – MeAIB | + 5 m*M* MeAIB |
| Normal medium | 100 | 97 ± 2.9 |
| Amino acid-free | 173 ± 4*** | 104 ± 6 |
| Normal medium + sucrose | 73 ± 4*** | 67 ± 1*** |
| Amino acid-free + sucrose | 232 ± 18*** | 82 ± 1*** |

Cells were incubated for 12 h in either normal medium or amino acid-free medium and then for a further 12 h in the presence or absence of 200 m*M* sucrose. Results are the means ± SEM of at least four replicate plates. ***$p < 0.001$ vs. control medium [data taken from 6].

Primers were made to the amino acid sequence IIEVEDWEIF spanning amino acids 270–279 and to YFNGGFAVDK spanning amino acids 503–512 of the published sequence of EAAC1 [7]. An internal probe was constructed using PCR, and the identity of the single 726-bp product was confirmed by sequencing. RNA was extracted from NBL-1 cells and Northern analysis carried out by standard procedures.

## Results

### *Regulation of $Na^+$-Dependent Neutral Amino Acid Transport*

NBL-1 cells have been shown to catalyse the $Na^+$-dependent transport of alanine. Alanine transport was competitively inhibited by glutamine, phenylalanine and leucine showing that all these amino acids are transported on a common transport system as in brush border membrane vesicles [4]. This transport system was not inhibited by excess concentration of N-methylaminoisobutyrate (N-MeAIB), a specific inhibitor of System A which occurs in most other cell types.

Table 1 shows that incubation of NBL-1 cell cultures in an amino acid-free medium caused an increase in the rate of alanine transport over a 24-hour period. The increased rate of transport was entirely inhibited by N-MeAIB [see 5]. The increase in transport rate was inhibited by cycloheximide and by tunicamycin but not by the inhibitor of microtubule function colcemid (not shown). Amino acid transport was further induced by incubation of NBL-1 cells in a medium made hyperosmotic with sucrose; mannitol had similar effects [6]. Stimulation occurred only when System A had been derepressed by

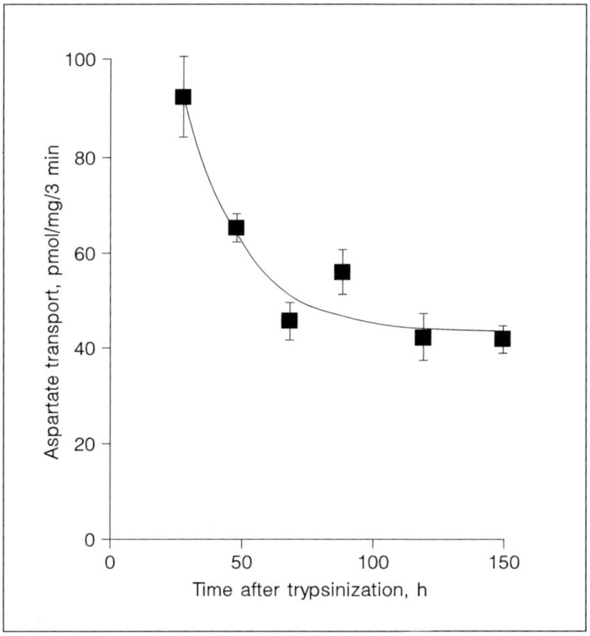


*Fig. 1.* Initial rate of $Na^+$-dependent *L*-aspartate (20 $\mu M$) uptake in NBL-1 cell cultures at various times after trypsinization.

prior incubation in an amino acid-free medium. The further induction by hyperosmotic medium was inhibited both by cycloheximide and colcemid. These results indicate that $Na^+$-dependent amino acid transport is stimulated in amino acid-free medium via derepression of System A but that the further induction of System A in hyperosmotic medium occurs via a different mechanism.

*$Na^+$-Dependent Glutamate Transport and Its Regulation*

Uptake of glutamate in NBL-1 cells was largely $Na^+$-dependent. *L*-Glutamate uptake was competitively inhibited by *L*-aspartate and *D*-aspartate but not *D*-glutamate. The $K_m$ value of the transporter for *L*-glutamate was 13 $\mu M$. These properties are similar to those of glutamate transport in bovine renal brush border membranes (not shown) and to System $X_{AG}$- which has been characterized in fibroblasts [3].

The activity of the glutamate transporter decreased with increasing cell density (fig. 1), reaching a plateau about 3–4 days after trypsinization. When cells at this stage of growth were incubated in amino acid-free medium, the

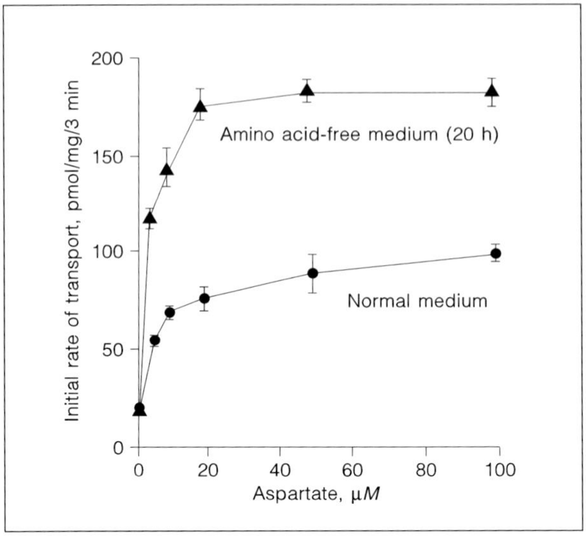


*Fig. 2.* Concentration dependence of *L*-aspartate transport in control or amino acid-starved cell cultures. The $Na^+$-dependent rates are presented.

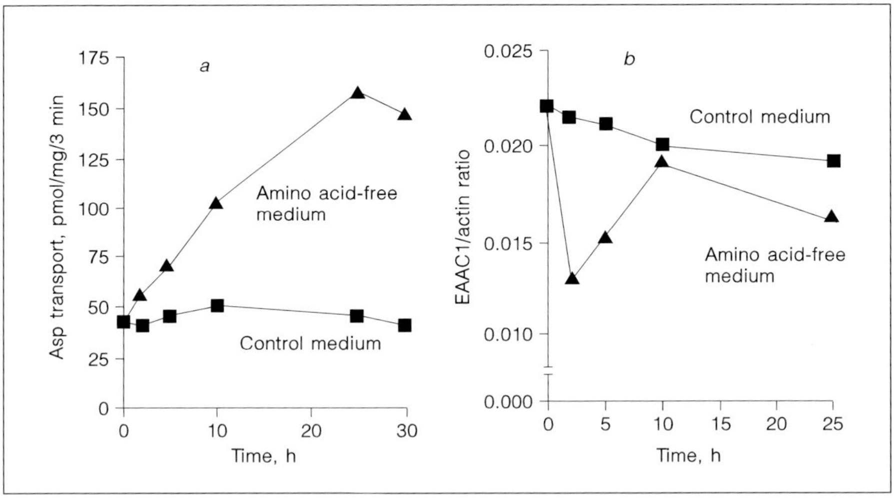


*Fig. 3.* *a* Initial rate of $Na^+$-dependent transport of 20 μ*M* aspartate in cell cultures starved of amino acids for various times. *b* Northern blot analysis. Total RNA was extracted from duplicate incubations of those in figure 3a. The RNA was fractionated and probed with the EAAC1 probe and also with a probe for actin to act as a loading control. Scans of the actin and EAAC1 autoradiograph signals were made and the ratio of the area of the EAAC1 signal to the actin signal is presented for each time point.

$V_{max}$ of the transport system increased twofold over 20 h while the $K_m$ was unchanged (fig. 2). The increase in activity was prevented by cycloheximide. The amino acid specificity of the increased rate of transport was the same as that of the basal rate, indicating that the increase was due to further synthesis of the same transport protein.

A high-affinity glutamate transporter (EAAC1) has been cloned from rabbit intestine [7]. The specificity and substrate $K_m$ values of the glutamate transporter in NBL-1 cells were similar to those of the EAAC1 transporter when expressed in Xenopus oocytes. An internal cDNA probe spanning amino acids 270–512 of the EAAC1 sequence was constructed by PCR. NBL-1 cells were incubated in normal or amino acid-free medium. The initial rate of aspartate transport was measured at intervals up to 30 h (fig. 3a). Total RNA was extracted from duplicate culture dishes and the EAAC1 probe used for Northern blot analysis (fig. 3b). The probe hybridized to a single 3.5-kb transcript in NBL-1 cell RNA. The level of the transcript declined slightly over 30 h in cells in normal medium and the rate of glutamate transport remained approximately constant. In cells incubated in amino acid-free medium the increase in the rate of glutamate transport was accompanied by an initial sharp decline in the level of the EAAC1 transcript followed by a recovery to control values. These results indicate that the primary regulation of the amount of functional EAAC1 glutamate transporter in NBL-1 cells occurs at the post-transcriptional level.

## Discussion

The above results demonstrate a number of regulatory mechanisms for amino acid transport in bovine renal epithelial cells. Amino acid starvation stimulates the synthesis of System A as in many other cell types. The induction of the high-affinity glutamate transporter by amino acid starvation has not been previously reported. The stimulation of previously derepressed System A activity by hyperosmolarity may involve the synthesis of a further regulatory protein since this is sensitive to colcemid while System A derepression itself is not.

Few studies have so far been undertaken to correlate level of transporter mRNA with transport activity. Induction of the EAAC1 is associated with a decrease in the level of EAAC1 mRNA. This suggests that the immediate increase in transport may result from increased translation of pre-existing mRNA or from some post-translational event. In contrast, the induction of the $Na^+$-dependent myoinositol transport system by exposure of MDCK cell to hyperosmotic medium is accompanied by greatly enhanced levels of specific mRNA for this transporter [8].

Stimulation of $Na^+$-amino acid cotransport would tend to increase cell volume. The mechanisms of regulation of transport described above may be of importance in the control of renal cell volume under conditions such as amino acid deprivation and exposure to hyperosmotic environments where cell volume would otherwise decrease.

## References


1 Lynch A, McGivan JD: Evidence for a single common $Na^+$-dependent transport system for alanine, glutamine, leucine and phenylalanine in brush border membrane vesicles from bovine kidney. Biochim Biophys Acta 1987;899:176–184.

2 Stevens BR: Amino acid transport in the intestine; in Kilberg MS, Haussinger D (eds): Mammalian Amino Acid Transport. New York, Plenum Press, 1992, pp 149–164.

3 Gazzola GC, Dall'Asta V, Bussolati O, Makowske M, Christensen HN: A stereoselective anomaly in dicarboxylic amino acid transport. J Biol Chem 1981;256:6054–6059.

4 Doyle FA, McGivan JD: The bovine renal cell line NBL-1 expresses a broad-specificity $Na^+$-dependent neutral amino acid transport system similar to that in bovine renal brush border membrane vesicles. Biochim Biophys Acta 1992;1104:55–62.

5 Felipe A, Soler C, McGivan JD: Amino acid deprivation leads to the synthesis of System A activity and the synthesis of a specific membrane glycoprotein in the bovine epithelial cell line NBL-1. Biochem J 1992;284:577–582.

6 Soler C, Felipe A, Casado FJ, McGivan JD, Pastor-Anglada M: Hyperosmolarity leads to an increase in derepressed System A activity in the renal epithelial cell line NBL-1. Biochem J 1993;289:653–658.

7 Kanai Y, Hediger MA: Primary structure and functional characterisation of a high-affinity glutamate transporter. Nature 1992;360:467–470.

8 Yamauchi A, Uchida S, Preston AS, Kwon HM, Handler JS: Hypertonicity stimulates transcription of gene for $Na^+$-myoinositol cotransporter in MDCK cells. Am J Physiol 1993;264:F20–F33.


J. D. McGivan, Department of Biochemistry, University of Bristol, Bristol BS8 1TD (UK)

Tizianello A, Baverel G, Endou H, Schoolwerth AC, O'Donovan DJ (eds):
Renal Ammoniagenesis and Interorgan Cooperation in Acid-Base Homeostasis.
Contrib Nephrol. Basel, Karger, 1994, vol 110, pp 109–114

# Hormonal and Dietary Control of Hepatic Glutamine Catabolism[1]

*John T. Brosnan*[a], *H. Stephen Ewart*[a], *Stephen A. Squires*[a], *Stephen H. Day*[a], *Zoran Kovacevic*[b], *Margaret E. Brosnan*[a]

[a] Department of Biochemistry, Memorial University of Newfoundland, St. John's, Nfld., Canada;
[b] Department of Biochemistry, Medical Faculty, Novi Sad, Serbia

Hepatic glutaminase plays a crucial role in glutamine homeostasis. Located on the inner membrane of mitochondria of periportal hepatocytes [1, 2], it is regulated in several ways. The total quantity of enzyme protein may vary as a result of changes in rates of synthesis and degradation, although such changes are relatively slow. The enzyme may be activated by ammonia [3] and, while the physiological function of this product activation is not clear, there is no question that it is effective at physiological ammonia concentrations. The enzyme is activated by inorganic phosphate, magnesium and bicarbonate [4].

In 1978, McGivan and co-workers [5, 6] made the important observation that hepatic glutaminase is regulated acutely by glucogenic hormones. In particular, exposure of hepatocytes to glucagon stimulates glutaminase. Mitochondria, isolated from animals injected 30 min previously with glucagon, exhibit increased flux through glutaminase. These observations are remarkable for a number of reasons. Alterations in total enzyme amount are not involved. The stimulation of flux through glutaminase is only evident in intact mitochondria; on disruption of the mitochondrial membranes the stimulation is lost. Finally, the effect is remarkably persistent; although mitochondria are isolated by routine procedures and no precautions are taken to preserve any possible covalent modification, the stimulated glutaminase is apparent even hours after mitochondrial isolation. In this respect this stimulation of glutaminase seems

[1] Supported by grants from the Medical Research Council of Canada to J. T. Brosnan and to M. E. Brosnan. Zoran Kovacevic was supported by an MRC East European Fellowship.

to be one of a family of persistent alterations in mitochondrial function brought about by glucagon [for review, see 7].

Glucagon injection is not a physiological event. However, ingestion of a high protein meal is associated with increased glucagon secretion. We therefore examined the effect of a single protein-rich meal on the action of hepatic glutaminase. We also report experiments on the signalling pathways involved in glucagon's activation of glutaminase and on the stimulatory effect of spermine on glutaminase.

## Materials and Methods

Male Sprague-Dawley rats, 250–350 g in weight, were used throughout. Mitochondria were prepared as described by Jois et al. [8]. Flux through glutaminase was measured by following the rate of glutamate production from glutamine, in the presence of rotenone and succinate, as described by Lacey et al. [6]. In experiments in which the effects of polyamines on glutaminase flux were examined, the various polyamines were preincubated with mitochondria for 6 min before the assay was started by the addition of substrate. Glutaminase activity in mitochondrial membranes was determined in freeze-thawed mitochondria as described by Szweda and Atkinson [9]. Isolated hepatocytes were prepared as described [10] except that hyaluronidase was omitted from the perfusate. Glutaminase flux was determined in hepatocytes by measuring $^{14}CO_2$ production from 1-$^{14}$C-glutamine [10]. Mitochondria were also prepared from isolated hepatocytes. The hepatocytes were incubated with various agents for 30 min and mitochondria were isolated by the procedure of Corvera and Garcia-Sainz [11].

## Results

*Stimulation of Hepatic Glutaminase by a Single High Protein Meal*

Rats were fed for 4 days on a diet containing 15% casein. They were then deprived of food for 6 h to ensure that they would promptly eat a meal when it was offered. The meal contained either 15 or 60% casein. When liver mitochondria were prepared 2 h later, the glutaminase flux was greatly enhanced in those from the rats fed the high protein meal (20.6 ± 1.7 vs. 7.1 ± 2.9 nmol/min/mg mitochondrial protein). In intact mitochondria from glucagon-treated rats the sensitivity of glutaminase to inorganic phosphate is known to be greatly increased [6]. Therefore, we examined the effect of varying the phosphate concentration of the incubation medium on the glutaminase flux of mitochondria isolated from rats fed the different meals. Figure 1 clearly demonstrates increased sensitivity of glutaminase flux to inorganic phosphate in mitochondria isolated from the rats fed the high protein meal. Fifty percent activation of flux occurred at 12 m*M* phosphate in rats fed the 15% casein meal versus 2 m*M* in those fed the high protein meal. The increased sensitivity to phosphate was not due to an in-

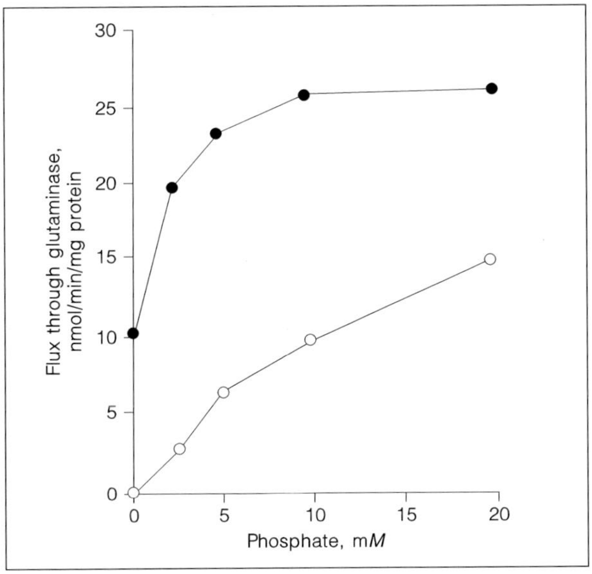


*Fig. 1.* Effect of inorganic phosphate on glutaminase flux in isolated mitochondria from rats fed a high (●) or low (○) protein meal.

creased matrix phosphate concentration. Intramitochondrial phosphate was measured after centrifugation of mitochondria through bromododecane and correction for extra-matrix phosphate by the distribution of $^{14}C$-sucrose. At an external phosphate concentration of 2.5 m*M* the matrix concentrations of phosphate were not different at 20.8 ± 1.4 and 17.2 ± 2.1 m*M* in mitochondria from rats fed the 60 and 15% protein meals, respectively.

When mitochondria were broken by freeze-thawing and glutaminase was assayed in the mitochondrial membranes we found no effect of the different meals, either on the total glutaminase activity or on its sensitivity to activation by inorganic phosphate. We have thus demonstrated in a physiological model the activation of glutaminase within intact mitochondria that is known to occur with glucagon injection.

*Signal Transduction and the Hormonal Activation of Hepatic Glutaminase*

We have investigated aspects of the signal transduction mechanism involved in the activation of glutaminase. The classical model of signal transduction by glucagon involves activation of adenyl cyclase giving rise to an increase in cyclic AMP which activates protein kinase A, which, in turn, phosphorylates a target protein. However, a second glucagon receptor, linked to the hydrolysis of inositol phospholipids, has also been described [12]. Therefore, we

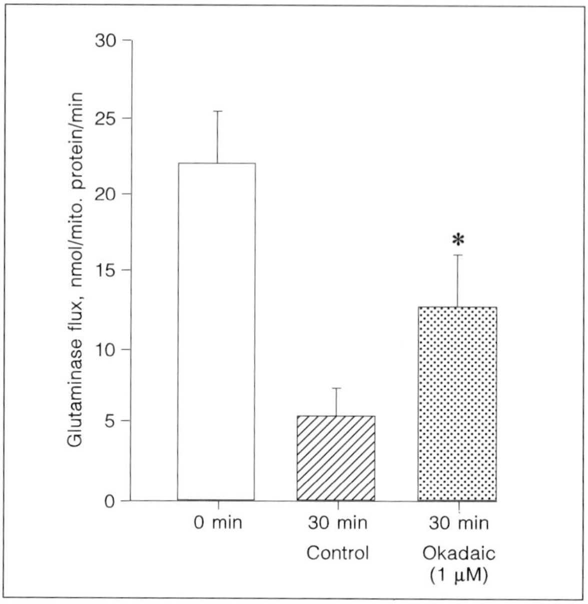


*Fig. 2.* Effect of okadaic acid (1 μ*M*) on glutaminase flux in mitochondria isolated from hepatocytes. Hepatocytes were incubated with okadaic acid or without addition for 30 min. Zero time hepatocytes were not incubated. Mitochondria, isolated from these hepatocytes, were used for measurement of glutaminase flux. $^{*}p < 0.05$ compared with the 30 min control incubations.

carried out a series of experiments to examine whether glutaminase can be activated by stimulation of protein kinase. In these experiments not only did glucagon stimulate glutaminase but also a variety of agents that elevated cyclic AMP were effective: forskolin, theophylline, dibutyryl cAMP. The protein kinase A activator, Sp-cAMPS [13] stimulated glutaminase in hepatocytes as did two agents, okadaic acid and calyculin A, whose known effect is to inhibit protein phosphatase 1 and 2A. These agents were not effective when incubated with isolated mitochondria. We also examined whether mitochondria prepared from these hepatocytes would display the 'persistent' stimulation of glutaminase that characterized those isolated after glucagon injection in vivo or ingestion of a high protein meal. Figure 2 shows the effect of incubation of hepatocytes with okadaic acid on glutaminase flux in subsequently isolated mitochondria. The relatively high glutaminase activity in mitochondria from unincubated hepatocytes decreased markedly upon incubation but this decrease could be partly attenuated by incubation with okadaic acid. The effects of glucagon were similar to that of the protein phosphatase inhibitor. Thus, the

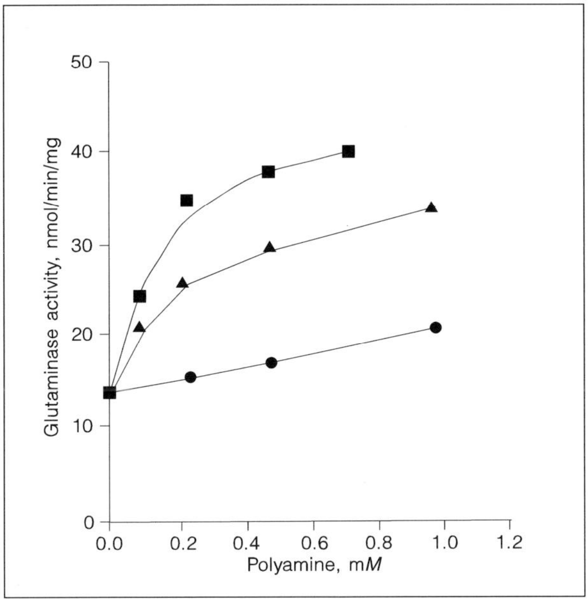


*Fig. 3.* Stimulation of glutaminase flux in isolated rat liver mitochondria by polyamines. ● = Putrescine; ▲ = spermidine; ■ = spermine.

effect of glucagon on the activation of glutaminase could be reproduced by an agent whose only known action is to inhibit protein phosphatase in the cytoplasm. We conclude that the activation of mitochondrial glutaminase involves the phosphorylation of a protein in the liver cytoplasm. We do not yet know the identity of the protein and the remaining details of signal transduction into the mitochondria remain obscure.

*Spermine as an Activator of Hepatic Glutaminase*

In the course of studies on the activation of hepatic glutaminase we discovered spermine to be a potent activator. In experiments with intact mitochondria preincubation with spermine, but not with putrescine or spermidine, led to marked stimulation of glutaminase. This was evident either as an increased rate of state 3 respiration of mitochondria with glutamine as sole substrate or as increased glutaminase flux (fig. 3) in intact mitochondria. The half-maximal effect of spermine was obtained at 0.25 m*M* which is in the physiological range. Further studies showed that spermine activation was evident in freeze-thawed mitochondria in which membrane-bound glutaminase is assayed in the absence of any permeability barriers. In this system, spermine was effective in the absence of any added ammonia. Spermine increased the enzyme's affinity for its substrate, decreasing the $K_m$ for glutamine from about 40 to 15 m*M* at a spermine concentration of 0.5 m*M*. It also made the enzyme markedly more sensitive to phosphate, decreasing the $M_{0.5}$ for phosphate activation from 15 to 4 m*M*.

These experiments suggest that spermine could be a physiologically significant activator of glutaminase. Spermine is known to be transported into energized mitochondria and the transport is dependent on the membrane potential [14]. Since glucagon administration *in vivo* increases the mitochondrial membrane potential, it is possible that an increased mitochondrial uptake of spermine could be a mechanism whereby glucagon activates glutaminase. This speculation is consistent with our findings that spermine markedly increases the affinity of glutaminase for inorganic phosphate, as does glucagon treatment. Further experiments are required to test this hypothesis fully.

## References


1 McGivan JD, Lacey JH, Joseph SK: Localization and some properties of phosphate-dependent glutaminase in disrupted liver mitochondria. Biochem J 1980;192:537–542.

2 Watford M, Smith EM: Distribution of hepatic glutaminase activity and mRNA in perivenous and periportal rat hepatocytes. Biochem J 1990;267:265–267.

3 Verhoeven AJ, Van Waarden JF, Joseph SK, Meijer AJ: Control of rat liver glutaminase by ammonia and pH. Eur J Biochem 1983;133:241–244.

4 Szweda LI, Atkinson DE: Response of rat liver glutaminase to pH, ammonia and citrate. Possible regulatory role of glutaminase in ureagenesis. J Biol Chem 1990;265:20869–20873.

5 Joseph SK, McGivan JD: The effect of ammonium chloride and glucagon on the metabolism of glutamine in isolated liver cells from starved rats. Biochim Biophys Acta 1978;543:16–28.

6 Lacey JH, Bradford NM, Joseph SK, McGivan JD: Increased activity of phosphate-dependent glutaminase in liver mitochondria as a result of glucagon treatment of rats. Biochem J 1981; 194:29–33.

7 Halestrap AP: Hormonal control of liver mitochondrial metabolism; in Kraus-Friedman N (ed): Hormonal Control of Gluconeogenesis. Boca Raton, CRC Press, 1986, vol 3, pp 31–48.

8 Jois M, Hall B, Fewer K, Brosnan JT: Regulation of hepatic glycine catabolism by glucagon. J Biol Chem 1989;264:3347–3351.

9 Szweda LI, Atkinson DE: Response of rat liver glutaminase to pH. J Biol Chem 1989;264: 15357–15360.

10 Krebs HA, Cornell NW, Lund P, Hems R: Isolated liver cells as experimental material; in Lundquist F, Tygstrup N (eds): Regulation of Hepatic Metabolism. Copenhagen, Munksgaard, 1974, pp 726–750.

11 Corvera S, Garcia-Sainz JA: Hormonal stimulation of mitochondrial glutaminase. Effects of vasopressin, angiotensin II and glucagon. Biochem J 1983;210:957–960.

12 Wakelam MJO, Murphy GJ, Hruby VJ, Houslay MD: Activation of two signal-transduction systems in hepatocytes by glucagon. Nature 1986;323:68–71.

13 Botelho LHP, Rothermel JD, Coombs RV, Jostosff B: cAMP analog antagonists of cAMP action. Methods Enzymol 1988;159:159–172.

14 Toninello A, DallaVia L, Siliprandi D, Garlid KD: Evidence that spermine, spermidine and putrescine are transported electrophoretically in mitochondria by a specific polyamine uniporter. J Biol Chem 1992;267:18393–18397.



Dr. John T. Brosnan, Department of Biochemistry, Memorial University of Newfoundland, St. John's, Nfld AlB 3X9 (Canada)

Tizianello A, Baverel G, Endou H, Schoolwerth AC, O'Donovan DJ (eds):
Renal Ammoniagenesis and Interorgan Cooperation in Acid-Base Homeostasis.
Contrib Nephrol. Basel, Karger, 1994, vol 110, pp 115–119

# Glutamine Production by Adipose Tissue in Metabolic Acidosis[1]

*Timothy Kowalski, Malcolm Watford*

Department of Nutritional Sciences, Thompson Hall, Cook College, Rutgers, The State University, New Brunswick, N. J., USA

Glutamine plays a major role as a transporter of nitrogen and carbon within the body, and is the major precursor of urinary ammonia during metabolic acidosis [1, 2]. In the rat the circulating level of glutamine decreases by 25–30% during chronic metabolic acidosis but this decrease is insufficient to provide for the extra glutamine utilization by the kidneys. In order to maintain glutamine homeostasis under such conditions, changes in extrarenal metabolism must occur. This could involve either increased production, or decreased utilization, of glutamine in a number of tissues.

Skeletal muscle has been long believed to be the major site of net glutamine synthesis in the body, with smaller contributions being made by the liver [1–4]. Both Goldstein [3] and Welbourne [1, 2] estimated the production and utilization of glutamine by different organs in the body and found total utilization to exceed total production. Recent reports show considerable glutamine release by the lungs [5–7], brain [8] and adipose tissue [9]. The purpose of the work reported here was to investigate the effect of acidosis on the release of glutamine by rat adipose tissue in vivo.

## Materials and Methods

Male Sprague-Dawley rats (350–400 g) were made acidotic by replacement of drinking water with 1.5% (w/v) ammonium chloride solution for 6 days. All animals had free access to a commercial rat diet (Purina). On the day of experiment, rats were anesthetized

[1] This work was supported by funds from the New Jersey Agricultural Experiment Station (Hatch project 14161) and the New York Obesity Research Center.

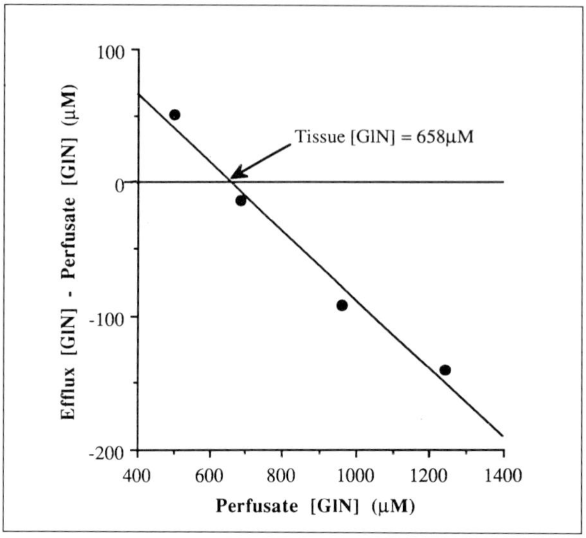


*Fig. 1.* Extracellular glutamine concentration in inguinal adipose tissue of the acidotic rat. A representative experiment is shown where a stepwise increase in the concentration of glutamine was perfused via a microdialysis probe located in an inguinal adipose depot of a 6-day acidotic rat. Extracellular glutamine concentration was obtained from the regression analysis ($y = 168.67–0.25629x$, $r = 0.98$). The arrow indicates the extracellular glutamine concentration (658 μ*M*), arterial glutamine concentration was 398 μ*M* resulting in a tissue-artery difference of 260 μ*M*.

with chloral hydrate (450 mg/kg b.w.) and microdialysis probes (0.2 × 30 mm, Cuprophan membrane MW cutoff 8–9 kD) inserted into an inguinal adipose depot. After a 30-min equilibration period, during which a solution of glutamine (500 μ*M*) was perfused, a stepwise increase in glutamine concentration (500–1, 200 μ*M*) was perfused. Determination of the efflux glutamine concentration and regression analysis (fig. 1) was used to calculate the extracellular glutamine concentration in the tissue. At the end of the experiment (1.25–1.5 h) arterial blood was drawn and assayed for glutamine content.

## Results

The arterial concentration of glutamine (table 1) was decreased by approximately 30% after 6 days' ammonium chloride acidosis, in keeping with previous reports [see 1, 4]. In control and acidotic animals, adipose tissue extracellular glutamine concentration was higher than in arterial blood indicating net glutamine production by this tissue (table 1, fig. 1). The difference was higher in acidotic animals (table 1).

*Table 1.* Glutamine production by rat inguinal adipose tissue

| | n | Glutamine concentration, μmol/l | | Difference |
|---|---|---|---|---|
| | | artery | adipose tissue | |
| Control | 3 | 666 ± 46 | 807 ± 46 | 140 ± 21 |
| Acidotic | 3 | 450 ± 35 | 728 ± 72 | 278 ± 41 |

Extracellular glutamine concentration in adipose tissue was calculated from the regression analysis of glutamine concentration in the microdialysis fluid as shown in figure 1. Results are means ± SEM for 3 separate rats in each group.

## Discussion

Adipose tissue possesses both the capacity for glutamine synthesis (glutamine synthetase) and glutamine hydrolysis (phosphate-activated glutaminase) and the direction of net flux in vivo has not been obvious [10]. This is due, in part, to problems in the study of adipose tissue metabolism in vivo. This tissue does not exist as a single mass with a clearly defined venous exit, and therefore arteriovenous difference measurements are not readily obtained. Experiments in vivo have shown net glutamine synthesis and net glutamine utilization, depending on the incubation conditions used [11–13]. Recently, Frayn et al. [9] developed an arteriovenous procedure for subcutaneous abdominal adipose tissue in humans. Using this technique they established adipose tissue as a site of net glutamine and alanine release, and of net glutamate uptake. However, this procedure is limited in its application and can only be used with this single adipose depot.

During the late 1980s a number of groups developed microdialysis techniques to study adipose tissue metabolism in vivo, but amino acid metabolism was not addressed [14, 15]. Using microdialysis procedure as developed by Lonnroth et al. [15] we have confirmed that the inguinal adipose depot is a site of glutamine and alanine production, and of glutamate uptake, in anesthetized rats [16] (see table 1). The precursors of adipose tissue glutamine synthesis have not yet been identified but likely candidates are glutamate, taken up by adipose tissue in vivo [9, 16], and branched chain amino acids, known to be catabolized by adipocytes in vitro [11].

Metabolic acidosis in the rat is accompanied by increased renal glutamine utilization. In order to maintain circulating glutamine levels there must be compensation by either increased production or decreased utilization in extrarenal tissues. In chronically acidotic rats there is increased glutamine production by skeletal muscle [1–3] and the liver changes from a site of net glutamine utilization to one of net glutamine production [1–3]. If the acidosis is accompanied by ketosis then the small intestine shows decreased glutamine utilization [17, 18], presumably substituting ketone bodies for glutamine as the major respiratory fuel. The potential for other tissues to provide additional glutamine for the kidneys in acidosis has not been established. The results presented here show that inguinal adipose tissue is a site of net glutamine synthesis and that increased adipose glutamine production may occur in metabolic acidosis. However, the model of acidosis used results in decreased food intake and a loss of adipose tissue mass, therefore without accurate estimations of blood flow and total adipose tissue mass it is not possible to draw any definitive conclusions. Similarly, different adipose tissue depots are known to exhibit differing rates of metabolism (e.g. higher rates of lactate production in intra-abdominal depots [19]) but to date, glutamine metabolism has only been studied in one depot and only in rats of one size (350–400 g). If increased glutamine production by adipose tissue does occur in the acidotic rat, this is not due to an adaptive change in glutamine synthetase activity [10]. The mechanism could involve changes in substrate availability, transporter activity, or alternatively, a decrease in glutamine hydrolysis by adipose tissue glutaminase. The net glutamine balance across adipose tissue is the difference between glutamine synthetase and glutaminase activities, but the actual flux through these two reactions is not known.

The results presented in this paper establish that adipose tissue is an important site of glutamine synthesis in both control and acidotic rats.

## References


1 Welbourne TC: Interorgan glutamine flow in metabolic acidosis. Am J Physiol 1987;253: F1069–F1076.

2 Welbourne TC, Joshi S: Interorgan glutamine metabolism during acidosis. J Parenter Enter Nutr 1990;14:77S–85S.

3 Schrock H, Goldstein L: Interorgan relationships for glutamine metabolism in normal and acidotic rats. Am J Physiol 1981;240:E519–E525.

4 Lemieux G, Watford M, Vinay P, Gougoux A: Metabolic changes in skeletal muscle during chronic metabolic acidosis. Int J Biochem 1980;12:75–83.

5 Welbourne TC: Role of the lung in glutamine homeostasis. Contrib Nephrol. Basel, Karger, 1988, vol 63, pp 178–182.

6 Souba WW: Lung glutamine metabolism. J Parenter Enter Nutr 1990;14:68S–70S.

7 Ardawi MSM: Glutamine-synthesizing activity in lungs of fed, starved, acidotic, diabetic, injured and septic rats. Biochem J 1990;270:829–832.

8 Grill V, Bjorkman O, Gutniak M, Lindquist M: Brain uptake and release of amino acids in nondiabetic and insulin-dependent diabetic subjects. Importance of glutamine release for nitrogen balance. Metabolism 1992;41:28–34.

9 Frayn KN, Khan K, Coppack S, Elia M: Amino acid metabolism in human subcutaneous adipose tissue. Clin Sci 1990;80:471–474.

10 Watford M: Regulation of expression of the genes for glutaminase and glutamine synthetase in the acidotic rat. Contrib Nephrol. Basel, Karger, 1991, vol 92, pp 211–217.
11 Tischler ME, Goldberg AL: Leucine degradation and release of glutamine and alanine by adipose tissue. J Biol Chem 1980;255:8074–8081.
12 Ardawi MSM, Jamal YS: Glutamine metabolism in skeletal muscle of glucocorticoid-treated rats. Clin Sci 1990;79:139–147.
13 Kowalchuk JM, Curi R, Newsholme EA: Glutamine metabolism in isolated incubated adipocytes in the rat. Biochem J 1988;249:705–708.
14 Watford M, Fried SK: Adipose tissue can now be directly studied in vivo. Trends Biochem Sci 1991;16:201–202.
15 Lonnroth P-A, Jansson P, Smith U: A microdialysis method allowing characterization of intercellular water space in humans. Am J Physiol 1957;253:E228–E231.
16 Kowalski T, Watford M: Glutamine metabolism in adipose tissue in vivo. FASEB J 1993;7: A387, abstr 2237.
17 Brosnan JT, Man KC, Hall DE, Colbourne SA, Brosnan ME: Interorgan metabolism of amino acids in streptozotocin-diabetic ketoacidotic rat. Am J Physiol 1983;244:E151–E158.
18 Watford M, Erbelding EJ, Smith EM: The regulation of glutamine and ketone-body metabolism in the small intestine of the long-term (40-day) streptozotocin-diabetic rat. Biochem J 1987;242:61–68.
19 DiGirolamo M, Newby FD, Lovejoy J: Lactate production in adipose tissue: A regulated function with extra-adipose implications. FASEB J 1992;6:2405–2412.

Malcolm Watford, DPhil, Department of Nutritional Sciences, Thompson Hall,
Cook College, Rutgers, The State University, POBox 231, New Brunswick, NJ 08903 (USA)

Tizianello A, Baverel G, Endou H, Schoolwerth AC, O'Donovan DJ (eds):
Renal Ammoniagenesis and Interorgan Cooperation in Acid-Base Homeostasis.
Contrib Nephrol. Basel, Karger, 1994, vol 110, pp 120–126

# $NH_3$ and $NH_4^+$ Permeance of the Colonocyte Apical Membrane

*Wolfgang E. Lohrmann*[a], *George M. Feldman*[a, b]

Departments of
[a] Medicine and [b] Physiology, McGuire Veterans Affairs Medical Center,
Medical College of Virginia/Virginia Commonwealth University, Richmond, Va., USA

In the human colon, 200–250 mmol of ammonia ($NH_3$ + $NH_4^+$) are absorbed daily [1, 2]. Studies of colonic function, performed in vivo, have suggested that diffusion of $NH_3$ is the dominant mode of absorption and that conversion of $NH_4^+$ to $NH_3$ by secreted base, $HCO_3^-$, generates $NH_3$ [3, 4]. However, data from small intestine, thick ascending limb of Henle's loop, and fish gill have suggested that $NH_4^+$ transport can be very important [5–11].

We, therefore, undertook studies to assess if diffusive ammonia flux occurs in an in vitro preparation of colon. Since the initial results were consistent with diffusive transport, we then evaluated the role $NH_3$ diffusion across the apical membrane of surface colonocytes. To do this we utilized the fact that $NH_3$ is a strong base, and measured changes in intracellular pH ($pH_i$) induced by ammonia solutions. Because the measured changes were a fraction of those predicted by a straightforward model of $NH_3$ diffusion, it appears likely that $NH_4^+$, an acid, also enters the cell and blunts the alkalinizing effect of $NH_3$. Moreover, it appears that in the colon $NH_4^+$ transport is quantitatively more important than $NH_3$ diffusion.

## Methods

*Tissue Preparation and Solutions*

Segments of descending colon were obtained from male Sprague-Dawley rats and stripped of serosal layers. The $HCO_3^-/CO_2$-buffered solution had the following composition (m*M*): 147 $Na^+$, 4.0 $K^+$, 24.6 $HCO_3^-$, 1.2 $Ca^{2+}$, 1.2 $Mg^{2+}$, 0.4 $H_2PO_4^-$, 1.6 $HPO_4^{2-}$, 10 glucose, and 128.8 $Cl^-$. When gassed with 5.6% $CO_2$ at 37 °C, this solution had a pH of 7.40 ± 0.03. HEPES (20 m*M*)-buffered solutions were nominally free of $HCO_3^-/CO_2$, but their cation composition was identical to that of the $HCO_3^-/CO_2$-buffered solution. When ammonia was in solution, it replaced $Na^+$.

*Measurement of Transepithelial Flux*

Transepithelial transport was measured in an Ussing-like chamber [12]. The mucosal surface was superfused at 0.1 ml/min, aliquots were taken from inflow and effluent, total ammonia concentration was measured by the Bertholet procedure [3], and ammonia ($NH_3 + NH_4^+$) flux was calculated.

*Measurement of $pH_i$*

$pH_i$ was measured in surface colonocytes loaded with 2'7'-bis(carboxyethyl)5–6-carboxyfluorescein (BCECF) [13, 14]. Colonic segments were mounted in a special assembly in a filter wheel fluorometer, and the mucosal and serosal surfaces were superfused independently. The procedure for rapidly exposing the surface cells to ammonia solutions was: stop sustaining superfusate fluid flow to the mucosal surface, aspirate mucosal chamber contents, inject new solution (>100 ml/min), aspirate contents, and restart sustaining superfusate (7 ml/min). Surface cells were loaded with BCECF by exposing the mucosal surface to 10 μ*M* BCECF-AM for 30 min. Excitation light alternated between 440 and 500 nm (bandwidth 10 nm) at 70 Hz while emitted light intensity was measured at 530 nm (bandwidth 10 nm). After subtracting autofluorescence, the ratio of emission intensities (F500/F440) was calculated and then filtered at 10 Hz. Two-point calibrations (pH 6.8 and 7.6) were performed at the end of each study with solutions containing 132 m*M* $K^+$, 20 m*M* $Na^+$ and 5 μg/ml nigericin.

## Results

*Flux Studies*

Unidirectional ammonia flux was measured in segments of stripped distal colon by superfusing one side of the tissue with 20 m*M* ammonia solutions; the other side was superfused with ammonia-free solution. Flux from mucosa to serosa was $2.8 \pm 0.3$ $\mu Eq \cdot h^{-1} \cdot cm^{-2}$ (n = 6) while flux from serosa to mucosa appeared to be slightly less $2.4 \pm 0.3$ $\mu Eq \cdot h^{-1} \cdot cm^{-2}$ (n = 6). These results suggested that there may be a small net flux in the absorptive direction. To assess net flux directly both tissue surfaces were superfused with solutions containing 20 m*M* ammonia. Net ammonia absorption occurred at $0.4 \pm 0.1$ $\mu Eq \cdot h^{-1} \cdot cm^{-2}$ (n = 6). Because the unidirectional fluxes were of similar magnitude and severalfold greater than net flux, it appeared that ammonia diffusion is the dominant mode of ammonia transport in the colon.

*$NH_3$ Diffusion across the Apical Membrane*

Since $NH_3$ readily diffuses across cell membranes [15, 16], $NH_3$ diffusion across the apical colonocyte membrane may be a particularly important component of colonic ammonia transport. To evaluate this possibility, we took ad-

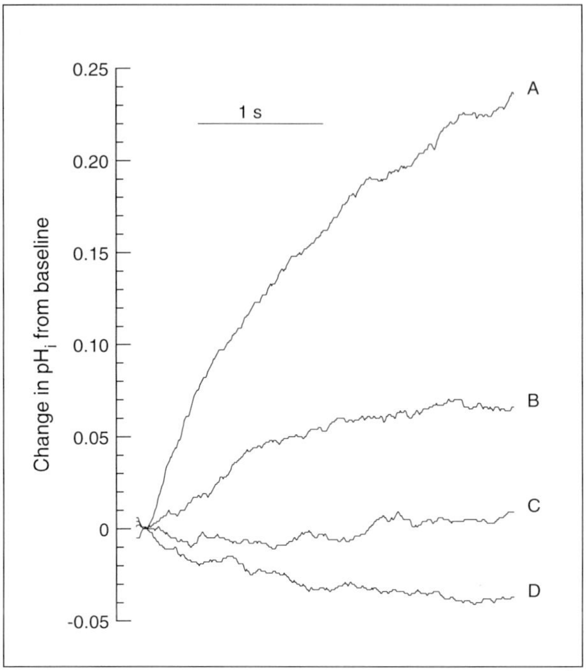


*Fig. 1.* Effect of $NH_3$ on $pH_i$. The mucosal surface of the colonocytes were repeatedly exposed to ammonia-containing solutions. In this representative experiment, trace A shows the effect of 25 m*M* ammonia-containing solution at pH 7.84 ($[NH_3] = 1.6$ m*M*), while trace B shows the effect at pH 6.22 ($[NH_3] = 0.04$ m*M*). Traces C and D show the effects of ammonia-free solutions at the same pHs as A and B, respectively. The baseline $pH_i$ values prior to each exposure were within 0.02 pH units of 7.33.

vantage of $NH_3$'s ability to alkalinize the cytoplasm. As shown in the Appendix (Model of $NH_3$ Diffusion), the rate of $pH_i$ change ($dpH_i/dt$) should vary directly with the $[NH_3]$ in the extracellular solution, and if $NH_3$ diffuses across the apical membrane while $NH_4^+$ does not cross the membrane, the relative change in $dpH_i/dt$ should equal the relative change in extracellular $[NH_3]$. Thus, we examined the hypothesis that only $NH_3$ diffuses across the apical colonocyte membrane by rapidly exposing the cell surface to varying concentrations of $[NH_3]$ and measuring $pH_i$ (fig. 1).

As shown in figure 2, relatively large increases in $[NH_3]$ failed to induce equivalent increases in $dpH_i/dt$. Indeed, the increment in $dpH_i/dt$ was just $5.9 \pm 1.4\%$ of the predicted increment ($p < 0.001$). This demonstrated that $NH_3$ diffusion alone is inadequate to account for the changes in $pH_i$ and raised the possibility that $NH_4^+$ also crosses the apical colonocyte membrane.

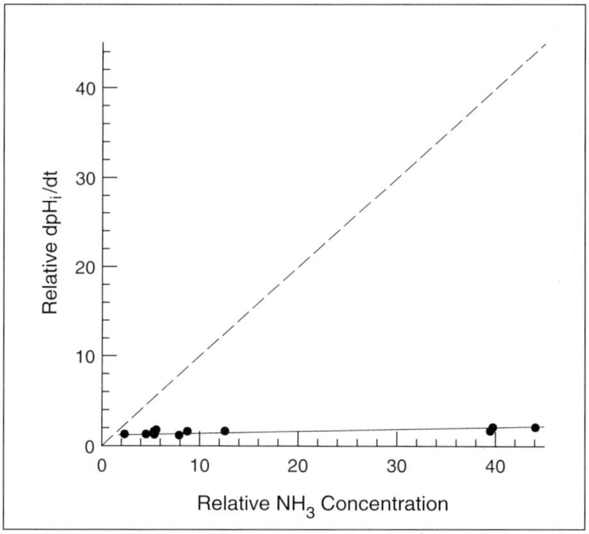


*Fig. 2.* Effect of varying [$NH_3$] on rate of change of intracellular pH. The dashed line shows the predicted effect, while the measured effects are shown by the filled circles.

## Discussion

In this study, the data from the transepithelial flux experiments were consistent with the notion that diffusion is the principle mode of ammonia transport in the colon, a conclusion that is similar to the overall conclusion of earlier studies [2–4]. Contrary to those earlier studies, however, the present study indicated that $NH_3$ diffusion is not the dominant mode of ammonia transport.

We found that $NH_3$ diffusion did not fully account for the $pH_i$ changes that occurred upon rapid exposure of colonocytes to ammonia-containing solutions. In fact, the measured rates of change in $pH_i$ were just 6% of those predicted by a straightforward model of $NH_3$ diffusion (fig. 2). The large discrepancy between the actual and the predicted results strongly suggests that an additional acid-base transport event(s) must have affected $pH_i$. Theoretically, one (or more) of the acid-base homeostatic mechanisms known to exist in colonocytes, such as $Na^+/H^+$ exchange, $Cl^-/HCO_3^-(OH^-)$ exchange, etc., could have blunted the increases in $pH_i$. However, these mechanisms do not affect $pH_i$ at very rapid rates, having maximum rates of approximately 0.2 pH units/min [13, 17]. On the other hand, the acidification rates necessary to blunt the $NH_3$ effects must be much greater, about 15 times greater than the observed alkalinization rates (observed range: 1 pH unit/min to 7 pH units/min).

Thus, it is likely that the above acid-base transport systems are not responsible for the discrepancy between the actual results and the predicted results.

Alternatively, movement of $NH_4^+$, an acid, could have affected $pH_i$, because $NH_4^+$ has been observed to acidify cells [10, 11]. Hence, $NH_4^+$ could have blunted the alkalinizing effect of $NH_3$, if $NH_4^+$ and $NH_3$ transited the colonocyte apical membrane simultaneously. To estimate the relative magnitudes of $NH_4^+$ and $NH_3$ fluxes, the apparent discrimination coefficient (i.e., ratio of apparent permeability coefficients) was calculated assuming that both $NH_4^+$ and $NH_3$ diffuse across the colonocyte apical membrane (see Appendix – Discriminaton Coefficient). Utilizing the data presented in figure 2 for the calculation, the discrimination coefficient was $2.4 \pm 0.6$ which suggests that $NH_4^+$ may be twice as permeant as $NH_3$. Moreover, since the $[NH_4^+]$ is far greater than the $[NH_3]$ at physiologic pH values, it is possible that 80 times more $NH_4^+$ than $NH_3$ crosses the apical membrane. Of course this estimation is based on the assumption that $NH_4^+$ flux is diffusive, and we did not evaluate the mechanism of $NH_4^+$ flux. In the thick ascending limb of Henle's loop, however, $NH_4^+$ has different kinetics, because it is transported by the $Na^+$-$K^+(NH_4^+)$-$2Cl^-$ cotransporter [6, 10, 11]. Nevertheless, the calculation suggests that in rat distal colon $NH_4^+$ flux is more important than previously thought. Indeed, the quantity of $NH_4^+$ crossing the apical membrane of surface colonocytes appears to be many times greater than quantity of $NH_3$.

## Appendix

*Model of $NH_3$ Diffusion*

The initial rate of $NH_3$ flux across a cell membrane can be expressed as:

$$J_{NH_3} = P_{NH_3} \cdot [NH_3], \qquad (1)$$

where $J_{NH_3}$ is the rate of $NH_3$ movement, $P_{NH_3}$ is the membrane permeability coefficient for $NH_3$, and $[NH_3]$ is the extracellular concentration. The initial intracellular $NH_3$ concentration is assumed to be negligible. Because $NH_3$ is a strong base, entry of $NH_3$ into the cell alkalinizes the cytoplasm. The rate of $pH_i$ change ($dpH_i/dt$) is a function of $J_{NH_3}$ and the buffering power ($\beta$) of the cytoplasm:

$$\frac{dpH_i}{dt} = \frac{J_{NH_3}}{\beta}. \qquad (2)$$

Combining these equations yields:

$$\frac{dpH_i}{dt} = \frac{P_{NH_3} \cdot [NH_3]}{\beta}. \qquad (3)$$

Assuming that $P_{NH_3}$ and β have constant values, changing the experimental $[NH_3]$ from $\alpha$ to $\beta$ changes $dpH_i/dt$ predictably:

$$\frac{(dpH_i/dt)_a}{(dpH_i/dt)_b} = \frac{[NH_3]_a}{[NH_3]_b}. \qquad (4)$$

This relationship was used to interpret experimental data. $P_{NH_3}$ was assumed to be constant, because comparisons were performed with data obtained from the same cells. β was assumed to be constant, because $pH_i$ was allowed to return to its initial value prior to each exposure to $NH_3$. Also, β did not change markedly during the measurement interval, because $dpH_i/dt$ was determined within the initial 1 s when $pH_i$ changes were modest.

*Discrimination Coefficient*

Since $NH_4^+$ is a weak acid, a portion of the $NH_4^+$ entering the cell donates $H^+$. The fraction donating $H^+$ is the fraction that generates $NH_3$ and is approximated by the term

$$\frac{10^{pH_i - pK_{amm}}}{1 + 10^{pH_i - pK_{amm}}}$$

or Φ, where $pK_{amm}$ is the pK for ammonia. This approximation is reasonable as long as $pH_i$ does not change markedly. Although $NH_3$ is a reasonably strong base, its potency is approximated by the term

$$\frac{1}{1 + 10^{pH_i - pK_{amm}}}$$

or γ. Hence, if it is assumed that both $NH_4^+$ and $NH_3$ diffuse into cells, then equation 3 can be modified:

$$\frac{dpH_i}{dt} = \frac{1}{\beta} \cdot \left[\gamma \cdot P_{NH_3} \cdot [NH_3] - \Phi \cdot P_{NH_4^+} \cdot [NH_4^+]\right]. \qquad (5)$$

Assuming that $P_{NH_3}$, $P_{NH_4^+}$ and β have constant values, changing the experimental $[NH_3]$ and $[NH_4^+]$ from *a* to *b* permits calculation of the discrimination coefficient or ratio of apparent permeability coefficients:

$$\frac{P_{NH_4^+}}{P_{NH_3}} = \frac{\left[\frac{dpH_i}{dt}\right]_a \cdot [NH_3]_b - \left[\frac{dpH_i}{dt}\right]_b \cdot [NH_3]_a}{\left[\frac{dpH_i}{dt}\right]_a \cdot [NH_4^+]_b - \left[\frac{dpH_i}{dt}\right]_b \cdot [NH_4^+]_a} \cdot 10^{pK_{amm} - pH_i}. \qquad (6)$$

## References


1 Walser M, Bodenlas L: Urea metabolism in man. J Clin Invest 1959;53:1617–1626.
2 Wrong OM: Nitrogen metabolism in the gut. Am J Clin Nutr 1978;31:1587–1593.
3 Cohen RM, Stephenson RL, Feldman GM: Bicarbonate secretion modulates ammonium absorption in rat distal colon in vivo. Am J Physiol 1988;254:F657–F667.
4 Swales JD, Tange JD, Wrong OM: The influence of pH, bicarbonate and hypertonicity on the absorption of ammonia from rat intestine. Clin Sci 1970;39:769–779.
5 Mossberg S, Ross G: Ammonia movement in the small intestine: Preferential transport by the ileum. J Clin Invest 1967;46:490–498.
6 Kinne R, Koenig B, Hannafin J, Kinne-Saffran E, Scott DK, Zeirold K: The use of membrane vesicles to study the NaCl/KCl cotransporter involved in active transepithelial chloride transport. Pflügers Arch 1985;405:101–105.
7 Goldstein L, Clairborne JB, Evans DE: Ammonia excretion by the gills of two marine teleost fish: The importance of $NH_4^+$ permeance. J Exp Zool 1982;219:395–397.
8 Koch MO, Hall MC: Mechanism of ammonium transport: Inhibition by potassium and barium. J Urol 1992;148:1285–1287.
9 Hall MC, Koch MO, McDougal WS: Mechanism of ammonium transport by intestinal segments following urinary diversion: Evidence for ionized $NH_4^+$ transport via $K^+$ pathways. J Urol 1992;148:453–457.
10 Kikeri D, Sun A, Zeidel ML, Hebert SC: Cellular $NH_4^+/K^+$ transport pathways in mouse medullary thick limb of Henle. Regulation by intracellular pH. J Gen Physiol 1992;99:435–461.
11 Kikeri D, Sun A, Zeidel ML, Hebert SC: Cell membranes impermeable to $NH_3$. Nature 1989;339:478–480.
12 Feldman GM, Berman SF, Stephenson RL: Bicarbonate secretion in rat distal colon in vitro: A measurement technique. Am J Physiol 1988;254:C383–C390.
13 King GG, Lohrmann WE, Feldman GM: Intracellular pH in surface colonocytes: Localization of $Na^+/H^+$ exchange to the apical membrane. Gastroenterology 1993;104:A257.
14 Kaunitz JD: Preparation and characterization of viable epithelial cells from rabbit distal colon. Am J Physiol 1988;254:G502–G512.
15 Boron WF: Intracellular pH transients in giant barnacle muscle fibers. Am J Physiol 1977; 233:C61–C73.
16 Boyarsky G, Ganz MB, Sterzel RB, Boron WF: pH regulation in single glomerular mesangial cells. I. Acid extrusion in absence and presence of $HCO_3^-$. Am J Physiol 1988;255:C844–C856.
17 Benya RV, Schmidt LN, Sahi J, Layden TJ, Rao MC: Isolation, characterization, and attachment of rabbit distal colon epithelial cells. Gastroenterology 1991;101:692–702.



George M. Feldman, MD, (111G), McGuire Veterans Affairs Medical Center,
1201 Broad Rock Road, Richmond, VA 23249 (USA)

Tizianello A, Baverel G, Endou H, Schoolwerth AC, O'Donovan DJ (eds):
Renal Ammoniagenesis and Interorgan Cooperation in Acid-Base Homeostasis.
Contrib Nephrol. Basel, Karger, 1994, vol 110, pp 127–132

# Time Course of Changes in mRNAs for Enzymes of Glutamine Metabolism in Kidney during Metabolic Acidosis[1]

*A.C. Schoolwerth, P. deBoer, A.F.M. Moorman, W.H. Lamers*

Anatomy and Embryology Laboratory, University of Amsterdam,
Academic Medical Center, Amsterdam, The Netherlands, and
Division of Nephrology, Medical College of Virginia,
Virginia Commonwealth University, Richmond, Va., USA

The classic study of Curthoys and Lowry [1] indicated that in response to $NH_4Cl$ acidosis, phosphate-dependent glutaminase (PDG) enzyme activity increased only in the proximal convoluted tubule. Subsequent studies by Good and Burg [2] demonstrated that this nephron segment was the source of increased ammonia synthesized in response to metabolic acidosis. It is known that in response to $NH_4Cl$ feeding, the activities of PDG, glutamate dehydrogenase (GDH) and phosphoenolpyruvate carboxykinase (PEPCK) increase over varying time courses [3]. The increase in PEPCK was the first noted and was due to increased enzyme synthesis [4]. The increase in PDG occurred more gradually, although recent studies by Hwang and Curthoys [5] suggested that PDG mRNA rose within 24 h. Glutamine synthetase (GS) was found to be localized exclusively to the pars recta of the proximal tubule [6] and, in response to acidosis, may decrease in activity.

To provide further insight into the regulation of renal ammoniagenesis, the changes in mRNAs for several enzymes important in glutamine metabolism were determined following induction of $NH_4Cl$ acidosis. Changes in enzyme appearance were evaluated by immunohistochemistry, while the changes in mRNAs were assessed by in situ hybridization and quantitated using the dot-blot technique.

[1] This work was supported in part by NIH grant DK-36822 and a Senior Fogarty Fellowship (FO6 TW01711). The authors are grateful to Dr. N. P. Curthoys for the generous gift of cDNA to rat renal PDG.

## Methods

Metabolic acidosis was induced in male Wistar rats by gavage feeding of 20 mmol $NH_4Cl$/kg b.w. This was the only feeding for animals sacrificed at 0, 2, 4, and 10 h. For animals killed 16, 30 and 72 h after initiation of $NH_4Cl$, maintenance doses of 5 mmol/kg b.w. were given twice daily. Animals sacrificed after 7 days were given 0.28 *M* $NH_4Cl$ as the sole source of drinking water after the initial gavage feeding. An additional group of animals was treated for 7 days with $NaHCO_3$ using a protocol similar to that for $NH_4Cl$, i.e. 20 mmol/kg b.w. as the initial dose followed by 0.28 *M* $NaHCO_3$ in the drinking water.

At designated times the animals were killed by cervical dislocation. Samples of whole kidney were fixed in a methanol, acetone, water mixture (40:40:20) for immunohistochemistry and in a solution containig 20 m*M* Hepes, pH 7.4/4% formaldehyde for in situ hybridization [7, 8]. Samples of liver, whole kidney, and kidney cortex were frozen for subsequent RNA isolation by the single-step acid guanidinium thiocyanate-phenol-chloroform extraction method [9]. Immunohistochemistry and Western analysis were performed as described previously [7]. mRNA measurements were performed utilizing cDNA for PDG (pGA 104) [10], GDH [11], PEPCK [12] and GS [13]. In situ hybridization was performed according to standard techniques utilized in the Anatomy and Embryology Laboratory using $^{35}S$-labeled probes [8]. mRNAs were quantitated by dot-blot hybridization utilizing $^{32}P$-labeled probes and a 108-bp SacII fragment of a human ribosomal RNA probe [14]. Approximately 40 animals were utilized in the study. Data analysis was by analysis of variance.

## Results

Changes in mRNA were measured in both whole kidney and in kidney cortex. While the changes were generally similar, they were more striking when evaluated in cortex alone. mRNA for PEPCK began to rise in kidney cortex at 4 h and was increased strikingly by 16-fold ($p < 0.05$) at 10 h after induction of acidosis; the levels decreased thereafter returning to control levels by 30 h. This surprising pattern was also seen for PDG and GDH. PDG mRNA was increased 2-fold at 4 h and by 10 h was increased 8-fold ($p < 0.05$) compared to control. As with PEPCK, PDG mRNA then decreased to levels equal to or below that of control between 30 h and 7 days. GDH mRNA increased less than the other two enzymes, but also peaked at 10 h ($p < 0.05$) after induction of acidosis, rising to a maximum of 2.6-fold that of the controls, before decreasing to control levels by 30 h.

GS mRNA changed very little, varying between 0.7- and 1.4-fold that of the control values. The lowest values were seen at 30 h (0.7-fold that of the control). Figure 1 summarizes the results of changes in PEPCK, PDG, GDH, and GS mRNAs at 10 and 30 h following initiation of $NH_4Cl$ feeding. Variable responses were seen following 7 days of $NaHCO_3$ feeding. PEPCK mRNA decreased significantly while GDH, GS, and PDG mRNAs were unchanged after 7 days of bicarbonate.

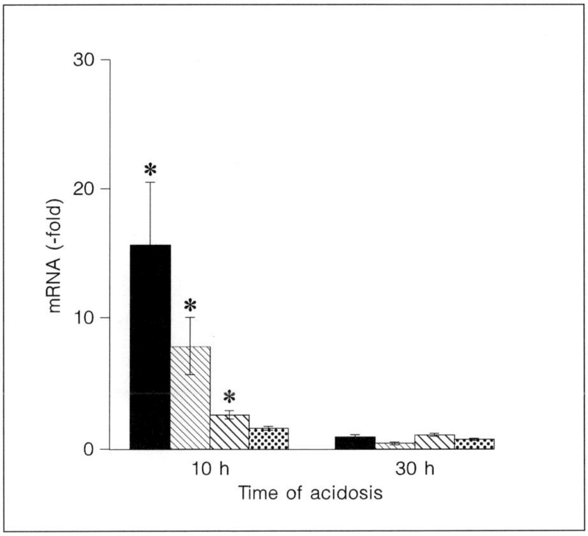


*Fig. 1.* Effect of $NH_4Cl$ acidosis on the relative levels of PEPCK (■), PDG (▧), GDH (▧), and GS (▩) mRNAs in kidney cortex. The increase in mRNAs at 10 and 30 h after initiation of acidosis is shown. Total RNAs from kidney cortex were quantitated by dot-blot analysis. All data were calculated as specific hybridization relative to that of ribosomol RNA and expressed as -fold change compared to control. The data are expressed as mean ± SE of values obtained from 4 to 6 rats. * $p < 0.05$ compared to controls.

Analysis of mRNAs by in situ hybridization demonstrated good agreement with the quantitative data obtained by dot-blot hybridization. Under control circumstances, in situ hybridization indicated that PEPCK mRNA was confined to the cortex and appeared to be localized to the medullary rays, consistent with localization of the enzyme in the $S_3$ segment of the proximal tubule. Following induction of acidosis, the pattern of PEPCK mRNA was visible throughout the cortex such that by 10 h virtually the entire superficial cortex contained message for PEPCK. As with the quantitative mRNA data, the PEPCK mRNA pattern decreased after reaching a maximum at 10 h, returning to the control pattern by 16–30 h.

The pattern of mRNA for PDG and GDH, as determined by in situ hybridization, demonstrated the presence of both mRNAs throughout a greater portion of the renal cortex 10 h following induction of $NH_4Cl$ acidosis compared to control. Thereafter, the patterns suggested a decrease in message approaching control levels by approximately 30 h.

Immunohistochemistry revealed increased but variable staining in kidney cortex for PEPCK, PDG, and GDH, which were confined largely to alkaline phosphatase staining tubules, suggesting localization primarily to proximal tubules. The increase in immunohistochemical staining for the enzymes occurred after the increase in mRNAs and, in contrast to in situ hybridization findings, were sustained. Thus, despite a decrease in mRNAs for all three enzymes, the amount of enzyme remained increased for up to 7 days. These increases were confirmed by Western analysis.

GS and GS mRNA were confined to the deep cortex and outer stripe of the medulla, but not superficial cortex, indicating localization of this enzyme in different renal cells than PDG and GDH. The staining pattern in deep cortex appeared to be in medullary rays, consistent with localization of the enzyme to the $S_3$ segment of the proximal tubule. This pattern did not change following $NH_4Cl$ feeding.

In contrast to the findings noted in kidney cortex, no substantial change in PEPCK, GDH, or GS was observed in liver by immunohistochemistry or by in situ hybridization. As quantitated by dot-blot hybridization, mRNAs for GDH and GS increased significantly at 4–10 h following administration of $NH_4Cl$, although the absolute changes were quite small (fig. 2).

## Discussion

These findings provide information on localization in kidney cortex of four enzymes and their mRNAs important in glutamine metabolism. Strikingly different patterns of enzyme localization were noted. Under physiological conditions, PEPCK staining was strongest in medullary rays but was also present in other alkaline phosphatase-positive cells, consistent with localization in the proximal tubule [6]. PEPCK mRNA was largely confined to medullary rays in control kidneys, suggesting that the $S_3$ segment of the proximal tubule contained the most PEPCK protein and mRNA. PDG and GDH demonstrated a more extensive localization in kidney cortex; although confined largely to the proximal tubule, activity was also found in occasional alkaline phosphatase-negative tubules. In contrast, GS was localized almost exclusively to deep cortex and outer medulla. Thus, this enzyme, formerly thought to be confined solely to the pars recta, apparently has a broader localization in the rat kidney.

Following induction of $NH_4Cl$ acidosis, mRNAs changed markedly. The most striking change occurred with PEPCK; increased PEPCK mRNA was found throughout proximal tubules and reached a maximum value by 10 h. These findings suggest that increasing amounts of PEPCK were synthesized in

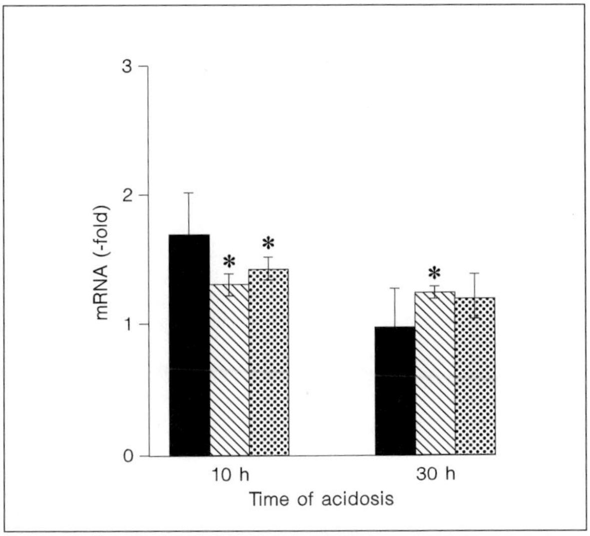


*Fig. 2.* Effect of $NH_4Cl$ acidosis on the relative levels of PEPCK (■), GDH (▨), and GS (▩) mRNAs in liver. The data are presented as described in figure 1. $*p < 0.05$ compared to controls.

renal cells which contained little, if any, enzyme under physiologic conditions. Similar, though less marked, increases were seen with PDG and GDH. Surprisingly, the mRNA for all three of these enzymes decreased after 10 h, and by 30 h had returned to control values. With respect to PDG, the levels actually decreased to below that of controls between 30 h and 7 days. These findings are quite different to those described by Hwang and Curthoys [5], who demonstrated an increase in PDG and PEPCK mRNA at 16 h which remained elevated; the reasons for the differences in these results are not apparent. Despite these changes in mRNA, increases in enzyme protein were sustained. GS staining and mRNA did not change.

These findings indicate that marked increases in PEPCK and PDG mRNAs occur during metabolic acidosis. Peak levels occur 10 h after initiation of acidosis, with mRNA then returning rapidly to control levels. This time course was also observed in the case of GDH, although the increase in mRNA was more modest in amount. Despite the subsequent decrease in mRNAs for these three enzymes after 16 h, a sustained increase in enzyme protein was noted.

## References


1 Curthoys NP, Lowry OH: Distribution of rat kidney glutaminase isoenzymes in the various structures of the nephron and their response to metabolic acidosis and alkalosis. J Biol Chem 1973;248:162–168.

2 Good DW, Burg MB: Ammonia production by individual segments of the rat nephron. J Clin Invest 1984;73:602–610.

3 Tannen RL: Renal ammonia production and excretion; in Windhager EE (ed): Handbook of Physiology. 8. Renal Physiology. New York, Oxford, 1992, pp 1017–1059.

4 Iynedjian PB, Ballard FJ, Hanson RW: The regulation of phosphoenolpyruvate carboxykinase synthesis in rat kidney cortex. J Biol Chem 1975;250:5596–5603.

5 Hwang JJ, Curthoys NP: Effect of acute alterations in acid-base balance on rat renal glutaminase and phosphoenolpyruvate carboxykinase gene expression. J Biol Chem 1991; 266:9392–9396.

6 Guder WG, Ross BD: Enzyme distribution along the nephron. Kidney Int 1984;26:101–111.

7 Lamers WH, Gaasbeek Janzen JW, Moorman AFM, Charles R, Knecht E, Martínez-Ramón A, Hernández-Yago J, Grisolía S: Immunohistochemical localization of glutamate dehydrogenase in rat liver: Plasticity of distribution during development and with hormone treatment. J Histochem Cytochem 1988;36:41–47.

8 Moorman AFM, deBoer PAJ, Das AT, Labruyère WT, Charles R, Lamers WH: Expression patterns of mRNAs for ammonia-metabolizing enzymes in the developing rat: The ontogenesis of hepatocyte heterogeneity. Histochem J 1990;22:457–468.

9 Chomczynski P, Sacchi N: Single-step method of RNA isolation by acid guanidinium thiocyanate-phenol-chloroform extraction. Anal Biochem 1987;162:156–159.

10 Shapiro RA, Farrell L, Srinivasan M, Curthoys NP: Isolation, characterization, and in vitro expression of a cDNA that encodes the kidney isoenzyme of the mitochondrial glutaminase. J Biol Chem 1991;266:18792–18796.

11 Das AT, Moerer P, Charles R, Moorman AFM, Lamers WH: Nucleotide sequences of rat liver glutamate dehydrogenase cDNA. Nucleic Acids Res 1989;17:2355.

12 Yoo-Warren H, Monahan JE, Short J, Short H, Bruzel A, Wynshaw-Boris A, Meisner HM, Samols D, Hanson RW: Isolation and characterization of the gene coding for cytosolic phosphoenolpyruvate carboxykinase from the rat. Proc Natl Acad Sci USA 1983;80:3656–3660.

13 deGroot CJ, ten Voorde GHJ, van Andel RE, te Kortschot A, Gaasbeek Janzen JW, Wilson RH, Moorman AFM, Charles R, Lamers WH: Reciprocal regulation of glutamine synthetase and carbamoylphosphate synthetase levels in rat liver. Biochim Biophys Acta 1987; 908:231–240.

14 Van Roon MA, Zonneveld D, Charles R, Lamers WH: Accumulation of carbamoylphosphate-synthetase and phosphoenolpyruvate-carboxykinase mRNA in embryonic rat hepatocytes. Eur J Biochem 1988;178:191–196.



Anton C. Schoolwerth, MD, Division of Nephrology, Department of Internal Medicine, Medical College of Virginia, Virginia Commonwealth University, Box 980 160, MCV Station, Richmond, VA 23298–0160 (USA)

Tizianello A, Baverel G, Endou H, Schoolwerth AC, O'Donovan DJ (eds):
Renal Ammoniagenesis and Interorgan Cooperation in Acid-Base Homeostasis.
Contrib Nephrol. Basel, Karger, 1994, vol 110, pp 133–137

# Acute Acidosis Inhibits Hepatic Amino Acid Uptake: Implications for Regulation of Acid-Base Balance

*L. Boon*[a], *P.J.E. Blommaart*[a], *A.J. Meijer*[b], *W.H. Lamers*[a], *A.C. Schoolwerth*[c,1]

[a] Department of Anatomy and Embryology, University of Amsterdam;
[b] E. C. Slater Institute, University of Amsterdam, The Netherlands;
[c] Department of Internal Medicine, Medical College of Virginia, Virginia Commonwealth University, Richmond, Va., USA

The traditional view of acid-base homeostasis has held that the kidney plays a primary role by maintaining the $HCO_3^-$ content in the extracellular fluid (ECF). Not only is filtered $HCO_3^-$ reabsorbed and returned to the ECF but the kidney synthesizes $HCO_3^-$ molecules consumed in buffering $H^+$ ions. The process of renal $HCO_3^-$ synthesis involves the metabolism of glutamine with the formation and excretion of $NH_4^+$ [1]. $NH_4^+$ excretion is critical because incorporation of ammonia into urea is $HCO_3^-$ consuming and this would counteract $HCO_3^-$ synthesis in the kidney.

This conventional interpretation has been challenged recently by the idea that the liver must be considered a primary organ of acid-base homeostasis. Proponents of this view have emphasized that large quantities of $HCO_3^-$ and $NH_4^+$ are formed as daily end products of amino acid metabolism [2]. $HCO_3^-$ accumulation is obviated by hepatic urea synthesis, which also incorporates the excess $NH_4^+$. Furthermore, the process of urea formation has been found to be pH sensitive such that at acid pH ureagenesis is decreased while glutamine production and release from the liver is increased [3]. In this way $HCO_3^-$ is spared to compensate for the fall in pH.

[1] The authors are grateful to George M. van Woerkom, Jan G. de Haan and Martinus A.W. Maas for their expert technical assistance. A.C.S. was supported by grants DK36822 and TWO1711.

Several mechanisms have been proposed by which the main $HCO_3^-$ utilizing process, the ornithine cycle, can be inhibited at low pH. One of the possible mechanisms by which urea synthesis is decreased at low pH is inhibition of hepatic glutamine degradation. Glutamine degradation is reduced at low pH not only because of lower glutaminase activity [4], but also because glutamine transport across the plasma membrane is decreased at low pH [5]. Other targets for pH control for urea synthesis have also been proposed. These include the degree of protonation of $NH_3$ (the true substrate for carbamoylphosphate synthase) [6], inhibition of carbonic anhydrase at low pH [7], and decreased synthesis at low pH of N-acetylglutamate [8], the essential activator of carbamoylphosphate synthase. Hepatic amino acid transport has been proposed by us to be the primary target for inhibition of ornithine cycle flux at low pH [9]. In order to obtain information on the contribution of these factors to the control of urea synthesis in vivo, we studied hepatic amino acid uptake and urea synthesis in rats during acute acidosis and alkalosis.

## Materials and Methods

Acute acidosis or alkalosis was induced in anesthesized (pentobarbital) adult female Wistar (200–250 g) rats by infusion of 0.15 *M* HCl or 0.15 *M* $NaHCO_3$, respectively, into the jugular vein at 4 ml/h for 3 h (total 1.8 mmol). Blood was obtained from the portal vein (the pH of which was also measured) and from the hepatic vein of the same animal. To ensure that blood was obtained from the hepatic vein, a clamp was placed across the inferior vena cava above the renal veins after portal blood had been sampled. A needle was then placed in the inferior vena cava, above the liver, with the tip directed towards the liver. The blood was quickly transferred into EGTA-coated tubes (0 °C). Amino acid analysis was carried out on plasma samples as described below; the detection limit was 5 nmol/ml plasma. After obtaining the plasma samples, the liver was quickly removed and frozen in liquid nitrogen. Subsequently, the liver was pulverized under liquid nitrogen and liver protein was precipitated with sulfosalicylic acid. Amino acids and ammonia were determined in the supernatant on a LKB alpha-plus amino acid analyzer. For liver samples, the detection limit of the amino acid analyzer was 0.5 nmol/mg protein.

Protein was determined using the enhanced protocol of the bicinchonninic acid (BCA) protein assay reagent (Pierce). DNA content per gram liver was determined according to Burton [10].

Protein from the urine was precipitated by addition of $HClO_4$ (final concentration 3%, m/v). After removal of the protein by centrifugation and neutralization of the supernatants to pH 7, ammonia and urea were determined according to Bernt and Bergmeyer [11].

Statistical significance of the difference between groups of observations was tested with Student's t test for paired observations.

*Table 1.* Portal venous blood pH in acute acidosis and alkalosis (results are means ± SD; n = 3)

| | | Acidosis | Alkalosis | p |
|---|---|---|---|---|
| Blood pH | | 7.12 ± 0.02 | 7.67 ± 0.03 | < 0.01 |
| Urinary excretion | $NH_4^+$ | 182 ± 30 | 27 ± 7 | < 0.01 |
| μmol N/h | Urea | 466 ± 21 | 780 ± 30 | < 0.01 |
| AV differences | $NH_4^+$ | –124 ± 21 | –130 ± 15 | NS |
| μ*M* | Alanine | –43 ± 31 | –173 ± 55 | < 0.05 |
| | Glycine | –73 ± 30 | –127 ± 14 | < 0.05 |
| | Glutamine | +44 ± 77 | +62 ± 66 | NS |
| | Total amino acids | –207 ± 100 | –612 ± 141 | < 0.05 |
| Central concentrations | $NH_4^+$ | 99 ± 28 | 50 ± 15 | < 0.05 |
| μ*M* | Alanine | 239 ± 37 | 88 ± 41 | < 0.05 |
| | Glycine | 175 ± 26 | 87 ± 14 | < 0.05 |
| | Glutamine | 534 ± 113 | 503 ± 186 | NS |
| | Total amino acids | 2,286 ± 79 | 1,402 ± 102 | < 0.05 |

– = Uptake; + = release, NS = not significant.

## Results

Portal venous blood pH was 7.1 and 7.7 in acute acidosis and alkalosis, respectively (table 1). Compared to alkalosis, urea synthesis decreased by 40% in acidosis, while renal ammonia excretion increased 7-fold (table 1). Net ammonia uptake by the liver was not different during acute acidosis and alkalosis. However, the ammonia concentration in portal blood was slightly higher and remained higher in central blood during acidosis as compared to alkalosis (table 1). Uptake of alanine and glycine, which are quantitatively the most important amino acids for degradation in the liver, was strikingly reduced at low blood pH as compared to high pH (75% reduction for alanine and 43% for glycine). Total hepatic amino acid uptake decreased by 66%, going from alkalosis to acidosis. Because of this, amino acid concentrations in the hepatic vein were higher in acidosis than in alkalosis, thereby increasing substrate availability for renal gluconeogenesis and amino acid oxidation. Hepatic vein alanine, for example, increased from 88 ± 41 μ*M* in alkalosis to 239 ± 37 μ*M* in acidosis. Surprisingly, no net glutamine release from the liver was found under either condition and the glutamine concentration in the hepatic vein, about 0.5 m*M*, was the same in acidosis and in alkalosis (table 1).

Intracellular branched chain amino acid concentrations were 2-fold higher in the liver in acidosis as compared to alkalosis (not shown). Because the branched chain amino acids are not catabolized in rat liver, this suggests an enhanced rate of hepatic proteolysis during acidosis. Indeed, the protein/DNA ratio decreased from $72 \pm 6$ in alkalosis to $61 \pm 1$ in acidosis. The intracellular concentrations of alanine and glycine were equal under both conditions. The intracellular concentrations of glutamate, aspartate and ammonia were relatively low during acidosis and increased during alkalosis. Of these, the 5-fold rise in glutamate, probably due to increased amino acid influx into the liver in alkalosis, followed by rapid transamination, was quantitatively the most important (not shown).

## Discussion

The relative contribution of liver and kidney in the maintenance of acid-base homeostasis has been the subject of considerable debate [1–3]. It is believed that decreased urea synthesis in acidosis favors hepatic synthesis of glutamine, which is then degraded in the kidney [3]. According to this hypothesis, the role of the liver in maintaining acid-base balance during acute acidosis involves periportal uptake of amino acids from the circulation with subsequent intracellular degradation of amino acids yielding equimolar amounts of $HCO_3^-$ and $NH_4^+$. Inhibition of urea synthesis during acidosis spares $HCO_3^-$ with the remaining $NH_4^+$ being incorporated into glutamine via glutamine synthase, localized in the hepatocytes surrounding the hepatic venules, prior to blood leaving the liver. This process would prevent $NH_4^+$ from entering the systemic circulation. If this is the correct mechanism for inhibition of urea synthesis, we would anticipate the following to occur during acidosis: (1) no effect on hepatic amino acid uptake; (2) increased hepatic glutamine production; (3) decreased urea production, and (4) increased $NH_4^+$ excretion in the urine. However, in contrast to these expectations, in acidosis we observed a strong decrease in hepatic amino acid consumption and no increased output of glutamine by the liver. Because the intracellular concentrations of readily transportable amino acids (alanine, glycine) were unchanged, we conclude that, in vivo, a decreased rate of transport of these amino acids across the hepatic plasma membrane rather than one of the steps in the ornithine cycle itself, primarily accounts for the decreased urea synthesis. This conclusion was supported by the observation that intracellular concentrations of glutamate and aspartate, the main products of amino acid degradation and substrates for urea synthesis, decreased during acidosis when urea synthesis is inhibited. Apparently, the elasticity coefficient, as defined in Westerhoff et al. [12], of the ornithine cycle for intracellular aspartate and glutamate is high. Whenever the elasticity coefficient of a process for its substrate is high, the flux control coefficient of this process, i.e. the ornithine cycle in this case, is low [12]. This indicates that the ornithine cycle can have only minor control during acute acidosis. In vitro experiments with isolated rat hepatocytes, incubated with physiological mixtures of amino acids at different pH values, led us previously to the same conclusion [9].

The increase in liver proteolysis in acidosis is reminiscent of the increase in muscle proteolysis observed under the same conditions [13]. This is an additional mechanism for the body to mobilize amino acids in order to increase $HCO_3^-$ production and to compensate for the fall in pH [14].

In summary, this in vivo study clarifies the role of renal and hepatic metabolism during acute acidosis. The liver is important because hepatic amino acid transport is decreased in acidosis. As a result, availability of amino acids for the kidney is increased. This increases renal amino acid metabolism with production of $HCO_3^-$ and excretion of $NH_4^+$. In this way, pH homeostasis is restored.

## References


1 Halperin ML, Goldstein MB, Stinebaugh BJ, Jungas L: Biochemistry and physiology of ammonia excretion; in Seldin DW, Giebisch G (eds): The Kidney: Physiology and Pathophysiology. New York, Raven Press, 1985, pp 1471–1490.
2 Atkinson DE, Camien M: The role of urea synthesis in the removal of metabolic bicarbonate and the regulation of blood pH. Curr Top Cell Regul 1982;21:261–302.
3 Häussinger D, Gerok W, Sies H: Hepatic role in pH regulation: Role of the intercellular glutamine cycle. Trends Biochem Sci 1984;9:300–302.
4 McGivan JD, Bradford NM, Verhoeven AJ, Meijer AJ: Liver glutaminase; in Häussinger D, Sies H (eds): Glutamine Metabolism in Mammalian Tissues. Berlin, Springer, 1984, pp 122–137.
5 Lenzen C, Soboll S, Sies H, Häussinger D: pH control of hepatic glutamine degradation. Role of transport. Eur J Biochem 1987;166:483–488.
6 Rémésy CH, Demigné C, Fafournoux P: Control of ammonia distribution ratio across the liver cell membrane and of ureogenesis by extracellular pH. Eur J Biochem 1986;158:283–288.
7 Häussinger D, Gerok W: Hepatic urea synthesis and pH regulation. Role of $CO_2$ and $HCO_3^-$, pH and the activity of carbonic anhydrase. Eur J Biochem 1985;152:381–386.
8 Kamemoto ES, Atkinson DE: Modulation of the activity of rat liver acetylglutamate synthase by pH and arginine concentration. Arch Biochem Biophys 1985;243:100–107.
9 Boon L, Meijer AJ: Control by pH of urea synthesis in isolated rat hepatocytes. Eur J Biochem 1988;172:465–469.
10 Burton K: A study of the conditions and mechanism of the diphenylamine reaction for the colorimetric estimation of deoxyribonucleic acid. Biochem J 1956;62:315–323.
11 Bernt E, Bergmeyer HU: Harnstoff; in Bergmeyer HU (ed): Methoden der enzymatischen Analyse. Weinheim, Verlag Chemie, 1970, pp 1738–1745.
12 Westerhoff HV, Groen AK, Wanders RJA: Modern theories of metabolic control and their applications. Biosci Rep 1984;4:1–22.
13 May RC, Kelly RA, Mitch WE: Metabolic acidosis stimulates protein degradation in rat muscle by a glucocorticoid-dependent mechanism. J Clin Invest 1986;77:614–621.
14 Meijer AJ, Lamers WH, Chamuleau RAFM: Nitrogen metabolism and ornithine cycle function. Physiol Rev 1990;70:701–748.



L. Boon, Academic Medical Centre, Department of Anatomy and Embryology,
Meibergdreef 15, NL–1105 AZ Amsterdam (The Netherlands)

Tizianello A, Baverel G, Endou H, Schoolwerth AC, O'Donovan DJ (eds):
Renal Ammoniagenesis and Interorgan Cooperation in Acid-Base Homeostasis.
Contrib Nephrol, Basel, Karger, vol 110, pp 138–143

# Effect of Chronic Acidosis on Hepatic Amino Acid Uptake and Gene Regulation: Implications for Control of Acid-Base Balance

*L. Boon*[a], *P.J.E. Blommaart*[a], *A.J. Meijer*[b], *W.H. Lamers*[a], *A.C. Schoolwerth*[c,1]

[a] Department of Anatomy and Embryology, University of Amsterdam;
[b] E. C. Slater Institute, University of Amsterdam, The Netherlands;
[c] Department of Internal Medicine, Medical College of Virginia, Virginia Commonwealth University, Richmond, Virginia, Va., USA

The traditional view of acid-base homeostasis holds that the kidney plays a primary role in buffering the $HCO_3^-$ content in the extracellular fluid (ECF). Filtered $HCO_3^-$ is not only reabsorbed and returned to the ECF, but the kidney also generates $HCO_3^-$ molecules to buffer $H^+$ ions. The process of renal $HCO_3^-$ synthesis involves the catabolism of glutamine with the formation and excretion of $NH_4^+$ [1]. $NH_4^+$ excretion is critical because incorporation of ammonia into urea is $HCO_3^-$ consuming and this would counteract $HCO_3^-$ synthesis in the kidney.

This conventional interpretation has been challenged by the concept that the liver must be considered a primary organ for the regulation of acid-base homeostasis. Proponents of this view have emphasized that large quantities of $HCO_3^-$ and $NH_4^+$ are daily formed as end products of amino acid metabolism [2]. $HCO_3^-$ accumulation is prevented by hepatic urea synthesis, which also incorporates the excess $NH_4^+$. The process of urea formation has been found to be pH-sensitive such that, at acid pH, ureagenesis is decreased. Furthermore, it has been proposed that glutamine production and release from the liver is increased [3]. In this way $HCO_3^-$ becomes available to compensate for the fall in pH.

[1] The authors are grateful to George M. van Woerkom, Jan G. de Haan and Martinus A.W. Maas for their expert technical assistance. A.C.S. was supported by NIH grants DK36822 and TW01711.

In the accompanying paper [4] we have shown that a decrease of amino acid transport across the hepatocyte plasma membrane is responsible for the decrease in urea synthesis in acute acidosis. In order to obtain information on the contribution of these effects under more physiological conditions, we measured urea synthesis, hepatic amino acid uptake and urinary excretion of urea and $NH_4^+$ as well as the levels of glutamine synthase (GS) and carbamoyl-phosphate synthase (CPS) mRNAs in liver and phosphoenolpyruvate carboxykinase (PEPCK), phosphate-dependent glutaminase (PDG), glutamate dehydrogenase (GDH), and GS mRNAs in kidney of rats during mild chronic acidosis and alkalosis.

## Materials and Methods

Chronic acidosis or alkalosis was induced in adult female Wistar (200–250 g) rats by addition of 40 mmol/kg/day of HCl or $NaHCO_3$, respectively, to food for 7 days. Blood was obtained from the portal vein (the pH of which was also measured) and from the hepatic vein of the same animal. To ensure that blood was obtained from the hepatic vein, a clamp was placed across the inferior vena cava above the renal veins after portal blood had been sampled. A needle was then placed in the inferior vena cava, above the liver, with the tip directed towards the liver. The blood was quickly transferred into EGTA-coated tubes (0 °C). Amino acid analysis was carried out on plasma samples as described below; the detection limit was 5 nmol/ml plasma. After obtaining the plasma samples, the liver was quickly removed and frozen in liquid nitrogen. Subsequently, the liver was pulverized under liquid nitrogen and liver protein was precipitated with sulfosalicylic acid. Amino acids and ammonia were determined in the supernatant on a LKB alpha-plus amino acid analyzer. For liver samples, the detection limit of the amino acid analyzer was 0.5 nmol/mg protein.

RNA was isolated by the guanidinium thiocyanate-phenol-chloroform method described by Chomczynski and Sacchi [5]. To verify that the RNA was not degraded, samples were analyzed by agarose gel electrophoresis under denaturing conditions followed by renaturation and staining with ethidium bromide, gels were exposed to ultraviolet light to visualize the bands resulting from the 28S and 18S ribosomal RNAs. These bands showed a ratio in fluorescence of 2 in intact RNA samples.

Relative GDH, GS, CPS, PEPCK and PDG mRNA concentrations were determined by quantitative dot-blot analysis using the respective cDNAs GDH [6], GS [7], CPS [8], PEPCK [9], PDG [10], and ribosomal RNA [11]. Approximately 2 µg RNA was diluted in 50 µl 10 m*M* Tris/HCl, 1 m*M* EDTA pH 8.0 for each sample. After addition of 0.6 volume 20 × SSC (1 × SSC: 150 m*M* NaCl, 15 m*M* Na-citrate pH 7.0) and 0.4 volume 37% formaldehyde, the samples were heated at 65 °C for 15 min and quenched on ice. After addition of 300 µl 10 × SSC, the samples were spotted on nitrocellulose membranes (BA-S 85, Schleicher & Schuell) using a dot-blot manifold (Bio-Rad). Membranes were prewetted by soaking for 5 min in $H_2O$ and for 15 min in 10 × SSC. Before and after loading the samples, the slots were washed with 400 µl 10 × SSC. After the final wash, the membranes were air-dried and incubated at 80 °C for 1 h to bind the RNA irreversibly to the membranes. Prior to hybridization the membranes were incubated in 40% (v/v) formamide, 0.9 *M* NaCl, 50 m*M* Na-phosphate pH 7.0, 5 m*M* EDTA, 0.1% (w/v) SDS, 5 × Denhardt's (1 × Denhardt's: 0.02% (w/v) Ficoll, 0.02% (w/v) polyvinylpyrrolidone, 0.02% (w/v) BSA), 100 µg/ml sheared herring testis DNA for 2 h at 42 °C. Hybridization was done at 42 °C for 16 h in the same solution supplemented with the probe and 10% dextran sulfate (for the hybridization with the ribosomal RNA gene probe the addition of dextran sulfate was omitted). After hybridization the membranes were sequentially washed in 3 × SSC, 0.1% (w/v) SDS; in 0.5 × SSC, 0.1% (w/v) SDS and in 0.1 × SSC, 1% (w/v) SDS, at 58 °C for 30 min each. Hybridization signals were quantified using a PhosphorImager™ (Molecular Dynamics).

Protein from the urine was precipitated by addition of $HClO_4$ (final concentration 3%, m/v). After removal of the protein by centrifugation and neutralization of the supernatants to pH 7, ammonia and urea were determined according to Bernt and Bergmeyer [12].

Statistical significance of the difference between groups of observations was tested with Student's t test for paired observations.

## Results

Portal venous blood pH was 7.31 and 7.41 in chronic acidosis and alkalosis, while the $HCO_3^-$ concentrations were 23.3 and 27.3 m*M*, respectively (table 1). Compared to alkalosis, urea synthesis decreased by 25% in acidosis, while renal ammonia excretion increased 25-fold (table 1). Net ammonia uptake by the liver was higher during chronic acidosis than alkalosis. The ammonia concentration in portal blood was higher in acidosis (292 ± 48 µ*M*) when compared to alkalosis (187 ± 37 µ*M*), while the ammonia concentration in hepatic venous blood was equal under both conditions (table 1). Uptake of alanine and glycine, which are quantitatively the most important amino acids for degradation in the liver, was reduced at low blood pH when compared to high pH (45% reduction for alanine and 33% for glycine). Total hepatic amino acid uptake decreased by 52%, going from alkalosis to acidosis, thereby increasing substrate availability for renal gluconeogenesis and amino acid oxidation. Surprisingly, glutamine release from the liver was equal under either condition, but the glutamine concentration in both the portal and the hepatic vein was higher in chronic alkalosis than in acidosis (table 1). Intracellular amino acid concentrations and the protein/DNA ratios in the liver were comparable in chronic acidosis and alkalosis (not shown).

In order to obtain more information on the induction of gene expression of some key enzymes involved in nitrogen metabolism in liver and kidney, we determined their relative concentrations after 7 days of chronic acidosis or alkalosis. In liver, CPS and GS mRNA were both 2-fold higher in the acidotic situation; in kidney, PEPCK and PDG mRNA were induced in acidosis while GDH and GS mRNA were not changed (table 2).

*Table 1.* Portal venous blood pH in chronic acidosis and alkalosis (results are means ± SD)

| | | Acidosis (n = 6) | Alkalosis (n = 4) | p |
|---|---|---|---|---|
| Blood pH | | 7.31 ± 0.02 | 7.41 ± 0.03 | < 0.01 |
| $[HCO_3^-]$, m*M* | | 23.3 ± 1.3 | 27.3 ± 1.4 | < 0.05 |
| Urinary excretion | $NH_4^+$ | 166 ± 22 | 6 ± 3 | < 0.01 |
| μmol N/h | Urea | 314 ± 36 | 421 ± 51 | < 0.01 |
| AV differences | $NH_4^+$ | –187 ± 39 | –58 ± 23 | NS |
| μM | Alanine | –110 ± 43 | –199 ± 67 | < 0.05 |
| | Glycine | –159 ± 30 | –238 ± 27 | < 0.05 |
| | Glutamine | +272 ± 77 | +325 ± 166 | NS |
| | Total amino acids | –190 ± 87 | –450 ± 124 | < 0.05 |
| Hepatic vein | $NH_4^+$ | 105 ± 24 | 129 ± 51 | NS |
| concentrations | Alanine | 440 ± 45 | 413 ± 49 | NS |
| μM | Glycine | 324 ± 37 | 241 ± 34 | < 0.05 |
| | Glutamine | 846 ± 71 | 1,122 ± 170 | < 0.05 |
| | Total amino acids | 2422 ± 51 | 1675 ± 125 | < 0.01 |

– = Uptake; + = release; NS = not significant.

*Table 2.* Gene expression of some key enzymes involved in nitrogen metabolism in liver and kidney (results are means ± SD)

| | | | Acidosis (n = 6) | Alkalosis (n = 4) | p |
|---|---|---|---|---|---|
| mRNA | Liver | CPS | 2.10 ± 0.79 | 1.15 ± 0.27 | < 0.05 |
| μg/μg rRNA | | GS | 3.48 ± 1.61 | 1.60 ± 0.16 | < 0.05 |
| | Kidney | PEPCK | 2.40 ± 0.97 | 0.94 ± 0.16 | < 0.05 |
| | | PDG | 1.22 ± 0.36 | 0.53 ± 0.17 | < 0.05 |
| | | GDH | 1.36 ± 0.33 | 1.29 ± 0.13 | NS |
| | | GS | 0.76 ± 0.21 | 0.71 ± 0.08 | NS |

## Discussion

The relative contribution of liver and kidney in the maintenance of acid-base homeostasis has been the subject of considerable debate [1–3]. It was suggested that decreased urea synthesis in acidosis favors hepatic synthesis of glutamine, which is then degraded in the kidney [3]. According to this hypothesis, the role of the liver in maintaining acid-base balance during acidosis involves periportal uptake of amino acids from the circulation with subsequent intracellular degradation of amino acids yielding equimolar amounts of $HCO_3^-$ and $NH_4^+$. Inhibition of urea synthesis during acidosis without effect on hepatic amino acid degradation spares $HCO_3^-$ with the remaining $NH_4^+$ being incorporated into glutamine via glutamine synthase, localized in the hepatocytes surrounding the hepatic venules, prior to blood leaving the liver. This process would prevent $NH_4^+$ from entering the systemic circulation. According to this hypothesis we would anticipate the following to occur during acidosis: (1) no effect on hepatic amino acid uptake; (2) increased hepatic glutamine production; (3) decreased urea production, and (4) increased $NH_4^+$ excretion in the urine. However, in contrast to these expectations, in both acute and chronic acidosis we observed a strong decrease in hepatic amino acid consumption and no increased output of glutamine by the liver. Because the intracellular concentrations of amino acids were unchanged, we conclude that, in vivo, a decreased rate of transport of these amino acids across the hepatic plasma membrane together with a decreased rate of amino acid degradation rather than one of the steps in the ornithine cycle itself, primarily accounts for the decreased urea synthesis. This conclusion is indirectly supported by the observation that the mRNA concentration of CPS, the key enzyme of the ornithine cycle, was increased despite the decrease in urea synthesis. These observations strongly suggest that intercellular glutamine cycling, which would spare large amounts of $HCO_3^-$ ions, as was proposed in Häussinger et al. [3], is very limited during both acute and chronic acidosis. Since decreased amino acid transport rates yield decreased rates of $HCO_3^-$ and $NH_4^+$ formation, urea synthesis is inhibited because of lack of substrates. Therefore, no net $HCO_3^-$ is spared in the liver, which is also supported by the observation that there is no extra output of glutamine or $NH_4^+$ by the liver during acidosis although GS mRNA was also induced in chronic acidosis (table 2).

Proponents of the view that $HCO_3^-$ generation in the kidney is of major importance during acidosis suggested that the renal induction of PDG and PEPCK mRNA during chronic acidosis (table 2) is necessary for increased renal gluconeogenesis from amino acids. However, since no net glutamine uptake by the kidney was observed during acidosis in the rat [13], the source of the 25-fold increase in $NH_4^+$ excretion in the urine remains unclarified. The results of this study suggest that the source of this $NH_4^+$ is the amino acids which were not degraded by the liver, rather than glutamine alone. In addition, both the mRNAs of GDH and GS, of which the respective enzymes are involved in glutamine metabolism, were not influenced during chronic acidosis (table 2).

In summary, this in vivo study clarifies the role of renal and hepatic metabolism during chronic acidosis. The liver is important because hepatic amino acid transport is decreased in acidosis. As a result, availability of amino acids to the kidney is increased. This increases renal amino acid metabolism with production of $HCO_3^-$ and excretion of $NH_4^+$. In this way, pH homeostasis is restored.

## References


1 Halperin ML, Goldstein MB, Stinebaugh BJ, Jungas L: Biochemistry and physiology of ammonia excretion; in Seldin DW, Giebisch G (ed): The Kidney: Physiology and Pathophysiology. New York, Raven Press, 1985, pp 1471–1490.

2 Atkinson DE, Camien M: The role of urea synthesis in the removal of metabolic bicarbonate and the regulation of blood pH. Curr Top Cell Regul 1982;21:261–302.

3 Häussinger D, Gerok W, Sies H: Hepatic role in pH regulation: Role of the intercellular glutamine cycle. Trends Biochem Sci 1984;9:300–302.

4 Boon L, Blommaart PJE, Meijer AJ, Lamers WH, Schoolwerth AC: Acute acidosis inhibits hepatic amino acid uptake: Implications for regulation of acid-base balance. Contrib Nephrol. Basel, Karger, 1994, vol 110, pp 133–137.

5 Chomczynski P, Sacchi N: Single-step method of RNA isolation by acid guanidinium thiocyanate-phenol-chloroform extraction. Anal Biochem 1987;162:156–159.

6 Das AT, Moerer P, Charles R, Moorman AFM, Lamers WH: Nucleotide sequence of rat liver glutamate dehydrogenase cDNA. Nucleic Acids Res 1989;17:2355.

7 Van de Zande L, Labruyère WT, Arnberg AC, Wilson RH, van den Bogaert AJW, Das AT, van Oorschot DAJ, Frijters C, Charles R, Moorman AFM, Lamers WH: Isolation and characterization of the rat glutamine synthetase-encoding gene. Gene 1990;87:225–232.

8 De Groot CJ, Zonneveld D, De Laaf RTM, Dingemanse MA, Mooren PG, Moorman AFM, Lamers WH, Charles R: Developmental and hormonal regulation of carbamoylphosphate synthethase gene expression in rat liver: Evidence for control mechanisms at different levels in the perinatal period. Biochim Biophys Acta 1986;866:61–70.

9 Yoo-Warren H, Monahan JE, Short J, Short H, Bruzel A, Wynshaw-Boris A, Meisner HM, Samols D, Hanson RW: Isolation and characterization of the gene encoding for cytosolic phosphoenolpyruvate carboxykinase from the rat. Proc Natl Acad Sci USA 1983;80:3656–3660.

10 Shapiro RA, Farell L, Srinivasan M, Curthoys NP: Isolation, characterization, and in vitro expression of a cDNA that encodes the kidney isoenzyme of the mitochondrial glutaminase. J Biol Chem 1991;266:18792–18796.

11 Laudien Gonzalez I, Gorski JL, Campen TJ, Dorney DJ, Erickson JM, Sylvester JE, Schmickel RD. Variation among human 28S ribosomal RNA genes. Proc Natl Acad Sci USA 1985; 82:7666–7670.

12 Bernt E, Bergmeyer HU: Harnstoff; in Bergmeyer HU (ed): Methoden der enzymatischen Analyse. Weinheim, Verlag Chemie, 1970, pp 1738–1745.

13 Squires EJ, Hall DE, Brosnan JT: Arteriovenous differences for amino acids and lactate across the kidney of normal and acidotic rats. Biochem J 1976;160:125–128.



L. Boon, Academic Medical Center, Department of Anatomy and Embryology,
Meibergdreef 15, NL–1105 AZ Amsterdam (The Netherlands)

Tizianello A, Baverel G, Endou H, Schoolwerth AC, O'Donovan DJ (eds):
Renal Ammoniagenesis and Interorgan Cooperation in Acid-Base Homeostasis.
Contrib Nephrol. Basel, Karger, 1994, vol 110, pp 144–149

# Renal Ammoniagenesis and Interorgan Flow of Glutamine in Chronic Metabolic Acidosis

*G. Deferrari, G. Garibotto, C. Robaudo, S. Saffioti, R. Russo, M.R. Sala, M. Bruzzone, A. Tizianello*

Department of Internal Medicine, Division of Nephrology,
University of Genoa, Italy

There is some evidence that in animals with chronic metabolic acidosis (CMA) muscle and splanchnic organs cooperate to supply the kidney with ammoniagenic precursors, mainly glutamine [1]. As for humans with CMA, investigations have provided insights into renal exchange of glutamine and other substrates [2–4], whereas the extrarenal metabolism of ammoniagenic precursors is poorly known. Moreover, in most studies performed so far, CMA has been induced by ammonium chloride, which could affect amino acid metabolism by providing a nitrogen load to the organs.

The study reported here has been undertaken in order to evaluate both renal ammoniagenesis and the exchange of ammoniagenic substrates across the kidney, splanchnic organs and peripheral tissues in man with CMA. In order to minimize the provision of N to the organs, CMA has been induced by a mixture of small aliquots of $NH_4Cl$ together with other acidifying agents.

## Material and Methods

*Patients*

Eight hypertensive patients (age 20–55 years) with normal renal function, were studied. Patients had arterial hypertension (MAP 135–157 mm Hg) and a renal vein catheterization had to be performed for measurement of plasma renin activity. Patients had normal routine laboratory tests and no history or evidence of gastrointestinal or hepatic diseases, congestive heart failure, diabetes mellitus or other endocrinopathies. The patients were eating a diet that provided 28–32 kcal, and 0.8–1 g of protein/kg/day. Medications were discontinued at least 6 days before the study. The patients were informed of the nature, purpose, procedure and possible risks before their voluntary consent was obtained. Procedures were in accordance with the Helsinki Declaration.

*Procedures*

The study was performed in the postabsorptive state. Six-day metabolic acidosis ($[HCO_3^-]_a$ 17±1 mmol/l) was induced in 4 out of 8 patients by oral HCl (1.2 mEq/kg), calcium chloride (0.6 mEq/kg) and $NH_4Cl$ (1.1 mEq/kg). The amount of $NH_4Cl$ given was 38% of the total acid load. The detailed procedures followed for these studies are reported elsewhere [3, 5, 6]. Briefly, arteriovenous differences across the kidney, splanchnic organs and leg for glutamine, other amino acids and ammonia were measured simultaneously in each patient. Glomerular filtration rate, blood flow across the kidney, splanchnic organs and leg and acid-base parameters in blood and urine were also obtained [3, 6].

*Analysis and Calculations*

Methods employed for the assay of glutamine, other amino acids, and ammonia, and for the measurement of GFR, blood flow across the kidney and splanchnic organs and acid-base parameters in blood and urine were extensively reported elsewhere [3, 6]. Leg blood flow was measured by $VO_2$ and leg $O_2$ uptake [7].

Total ammonia production was obtained by summing urinary ammonia and ammonia added to renal veins. The rate of uptake or release of metabolites by the organs was calculated by the arteriovenous difference multiplied by blood flow [3]. Standard statistical calculations have been employed [8]. Values are given as mean ± SEM.

## Results and Discussion

In CMA (fig. 1) urinary ammonia excretion markedly increases, while ammonia added to the renal vein diminishes slightly. Total ammonia production increases by 83%. Also glutamine extraction by the kidney is augmented. Glutamine uptake by the kidney is related to the degree of metabolic acidosis (fig. 2): the lower the arterial $[HCO_3^-]$, the higher the renal uptake of glutamine. The ratio of ammonia production to glutamine N extraction is 0.96 (fig. 1). Even if, according to this value, N supplied by glutamine may account for total ammonia production, a ratio very close to 1 suggests that other substrates besides glutamine are utilized for ammonia production. As a matter of fact, the renal uptake of glycine (9±1 vs. 5±3 μmol/min · 1.73 $m^2$ in controls; $p < 0.05$) increases in CMA.

In CMA a tendency towards an increased uptake and reduced release of amino acids by the kidney is observed. This finding appears clearly when the balance of N contributed by amino acids across the kidney is evaluated. In acidotic subjects the N in extracted amino acids is 24% increased (118±3 vs. 95±4 μmol/min · 1.73 $m^2$ in controls; $p < 0.05$) while the N in released amino acids is 48% diminished (48±5 vs. 93±6 μmol/min · 1.73 $m^2$ in controls;

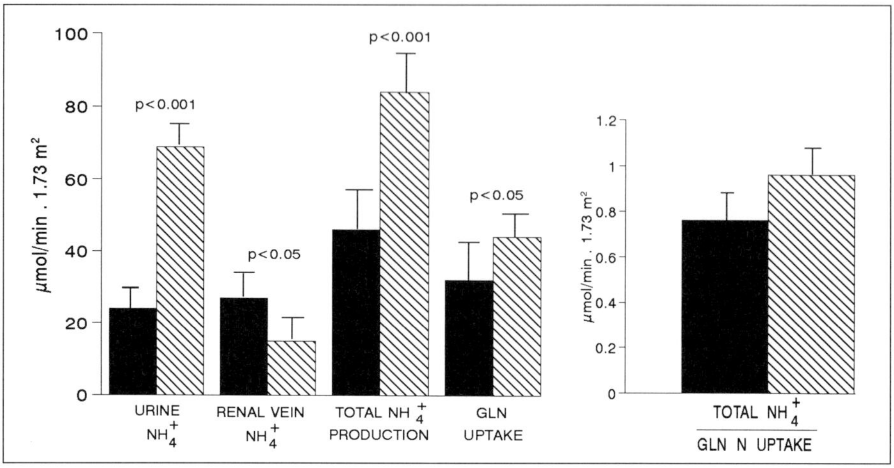


*Fig. 1.* Renal ammoniagenesis in CMA (▧) and controls (■).

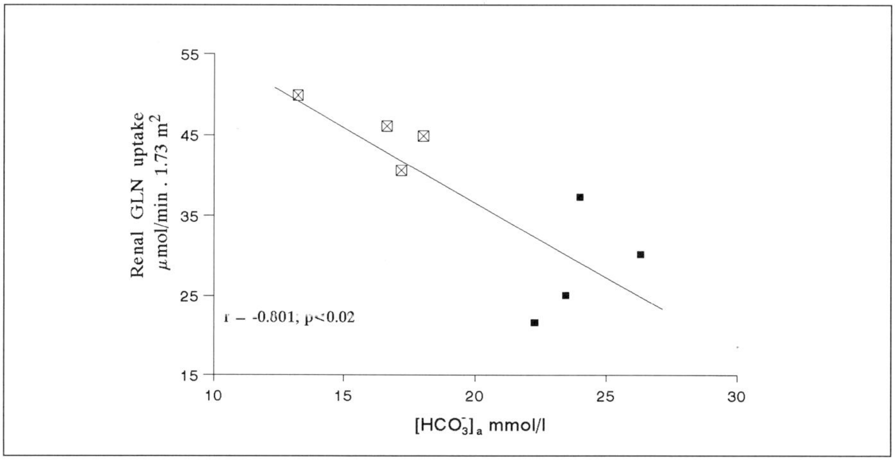


*Fig. 2.* Relationship between renal glutamine uptake and $[HCO_3^-]_a$ in CMA (⊠) and controls (■).

$p < 0.05$). As a consequence of these changes, renal balance of N contributed by amino acids is markedly positive (69 ± 11 vs. 2 ± 14 µmol/min · 1.73 $m^2$ in controls; $p < 0.05$). However, taking into account also the N contributed by total ammonia production, N balance is almost neutral. Accordingly, in man, metabolic acidosis markedly affects renal exchange not only of glutamine but also of other amino acids.

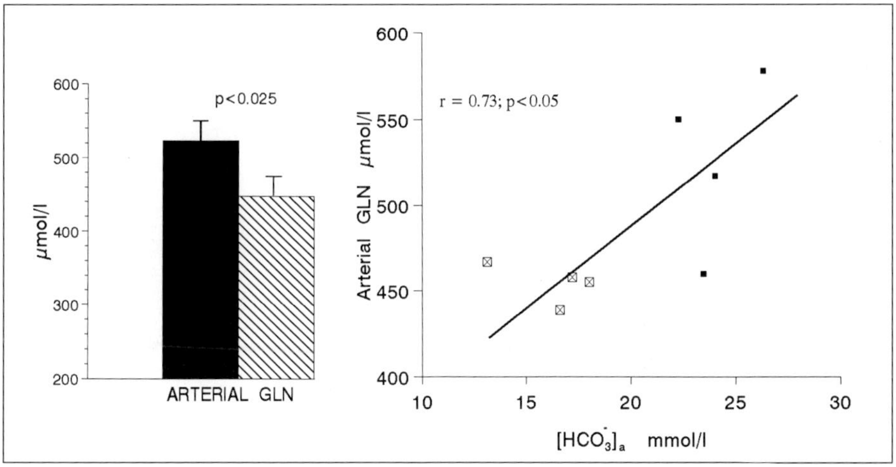


*Fig. 3.* Arterial levels of glutamine and relationship between arterial glutamine and $[HCO_3^-]_a$ in CMA (▧, ⊠) and controls (■, ■).

In CMA, blood glutamine levels decline and are directly correlated with arterial bicarbonate (fig. 3). So, glutamine is taken up by the kidney in increased amounts even in the presence of decreased blood glutamine levels. Possible adaptative changes in the provision of glutamine from other organs are not able to maintain the homeostasis of glutamine levels and supply to the kidney.

In normal postabsorptive subjects, splanchnic organs extract large amounts of glutamine and other amino acids from arterial blood [6, 9]. Figure 4 depicts splanchnic uptake of glutamine, alanine and N contributed by total amino acids. In CMA, glutamine uptake is slightly but not significantly decreased (–27%), and the uptake of alanine as well as total amino acid uptake are unchanged. Accordingly, these preliminary data suggest that in man, splanchnic organs fail to cooperate with the kidney by increasing glutamine supply to the kidney. Conversely, data obtained in rats with chronic $NH_4Cl$ acidosis [1, 10] show a decreased uptake of glutamine by the liver; again in rats the uptake of glutamine by the intestines seems to be reduced [1]. Apart from species differences, the experimental protocol for CMA induction used in the present study could explain these differences.

In the postabsorptive state, peripheral tissues, mainly muscle, are the main site for glutamine production and release into the circulation [9]. In CMA, the release of glutamine from the leg tends to be increased ($37 \pm 5$ vs. $26 \pm 4$ µmol/min · 1.73 $m^2$ in controls; $p < 0.075$). At the same time the release of branched-chain amino acids (BCAA) observed in

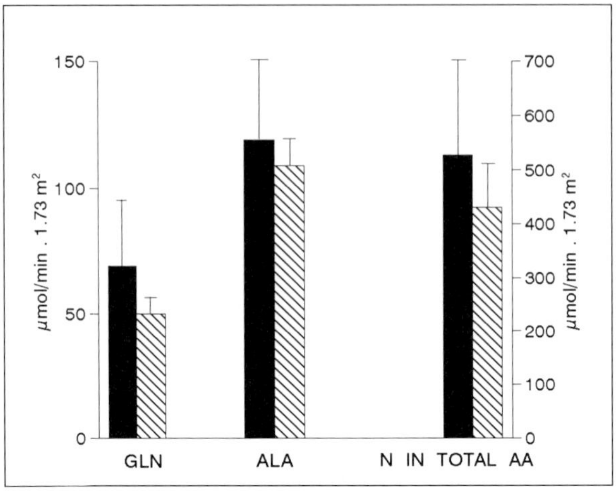


*Fig. 4.* Splanchnic exchange of glutamine, alanine and N contributed by total amino acids in CMA (▧) and controls (■).

normal acid-base balance is reversed into a small uptake (+4.0 ± 2 vs. –11.8 ± 3 μmol/min · 1.73 $m^2$ in controls; $p < 0.05$). Total amino acid release is unchanged (123 ± 10 vs. 110 ± 8 μmol/min · 1.73 $m^2$ in controls). These data suggest that peripheral tissues cope with the increased renal needs for glutamine with an increased supply of this substrate. Accordingly, in CMA an increased muscle BCAA degradation likely provides N for the augmented glutamine synthesis. It has been shown that in rats metabolic acidosis increases muscle BCAA degradation by stimulating the activity of branched-chain ketoacid dehydrogenase [11].

In conclusion, in man with 6-day metabolic acidosis induced by acid load given with a minimal dose of $NH_4Cl$, glutamine extracted by the kidney is augmented, even in the presence of decreased blood glutamine levels. Amino acids other than glutamine, mainly glycine, likely contribute to increased ammoniagenesis. The experimental setting used here does not show a significant cooperative role of splanchnic organs in glutamine provision to the kidney. The augmented renal demand for glutamine is met by peripheral tissues through an increased release of the same amino acid. A de novo synthesis of glutamine from BCAA in peripheral tissues is likely a source for glutamine release. These preliminary data indicate that adaptative changes in extrarenal glutamine metabolism are only partially able to maintain the homeostasis of circulating glutamine levels and supply to the kidney.

## Acknowledgements

This study was supported by grants from the Consiglio Nazionale delle Ricerche under the cooperative agreement between Italy and Canada, and by grants from the Consiglio Nazionale delle Ricerche, Progetto Finalizzato Invecchiamento, Sottoprogetto 3 and the Ministero dell'Università della Ricerca Scientifica e Tecnologica (Assegnazione per la ricerca scientifica 40 e 60%).

The authors wish to thank Mr. M. Marchelli and Ms. C. Sabatino for their excellent technical assistance and Mrs. F. Tincani for preparation of the manuscript and the figures.

## References

1 Welbourne TC: Interorgan glutamine flow in metabolic acidosis. Am J Physiol 1987;253: F1069–F1076.
2 Owen EE, Robinson RR: Amino acid extraction and ammonia metabolism by the human kidney during the prolonged administration of ammonium chloride. J Clin Invest 1963;42:263–276.
3 Tizianello A, Deferrari G, Garibotto G, Robaudo C, Acquarone N, Ghiggeri GM: Renal ammoniagenesis in an early stage of metabolic acidosis in man. J Clin Invest 1982;69:240–250.
4 Tizianello A, Deferrari G, Garibotto G, Robaudo C, Bruzzone M, Passerone GC: Renal ammoniagenesis during the adaptation to metabolic acidosis in man. Contrib Nephrol. Basel, Karger, 1982, vol 31, pp 40–46.
5 Deferrari G, Garibotto G, Robaudo C, Canepa A, Bagnasco S, Tizianello A: Leg metabolism of amino acids and ammonia in patients with chronic renal failure. Clin Sci 1985;69:143–151.
6 Deferrari G, Garibotto G, Robaudo C, Sala MR, Tizianello A: Splanchnic exchange of amino acids after amino acid ingestion in patients with chronic renal failure. Am J Clin Nutr 1988; 48:72–83.
7 Jorfeldt L, Wahren J: Leg blood flow during exercise in man. Clin Sci 1971;41:459–473.
8 Snedecor GW, Cochran WG: Statistical Methods, ed 6. Ames, Iowa State University Press, 1967.
9 Felig P: Amino acid metabolism in man. Annu Rev Biochem 1975;44:933–939.
10 Boon L, Blommaart PJE, Meijer AJ, Lamers WH, Schoolwerth AC: Effect of chronic acidosis on hepatic amino acid uptake and gene regulation: Implications for control of acid-base balance. Contrib Nephrol. Basel, Karger, 1994, vol 110, pp 138–143.
11 May RC, Hara Y, Kelly RA, Block KP, Buse MG, Mitch WE: Branched-chain amino acid metabolism in rat muscle: Abnormal regulation in acidosis. Am J Physiol 1987;252:E712–E718.

Giacomo Deferrari, MD, Dipartimento di Medicina Interna, Divisione di Nefrologia,
Viale Benedetto XV, 6, I–16132 Genova (Italy)

Tizianello A, Baverel G, Endou H, Schoolwerth AC, O'Donovan DJ (eds):
Renal Ammoniagenesis and Interorgan Cooperation in Acid-Base Homeostasis.
Contrib Nephrol. Basel, Karger, 1994, vol 110, pp 150–157

# Effects of Benzoate and Hippurate on the Regulation of Interorgan Nitrogen Flow: Therapeutic Benefits[1]

*P.D. Dass*[a], *B.P. Stuart*[a], *E.W. Holmes*[b]

[a] Miles Inc., Agriculture Division, Toxicology Department, Stilwell, Kans.;
[b] Department of Pathology/Clinical Chemistry, Loyola University Medical Center, Maywood, Ill., USA

Blood levels of the major nitrogen-carrying metabolites glutamine, glutamate, urea, and glutathione are under interorgan regulatory control, and reflect homeostatic mechanisms that operate at several levels [1–3]. The mechanisms include the regulation of fluxes through the γ-glutamyl cycle [4], the urea cycle, and the tricarboxylic acid cycle, the net affects of which determine interorgan fluxes of substrates and metabolites. This study examines the intraorgan metabolism and the interorgan flow of major nitrogen-carrying metabolites across the liver, gut and kidney. These studies provide a rationale for the therapeutic benefits of redirecting metabolite flow using physiological modulators such as hippurate, and its synthetic precursor, benzoate.

## The Problem

Recently, benzoate has been used clinically for the treatment of hyperammonemia and other disorders of nitrogen metabolism [5, 6]. Benzoate is conjugated with glycine in the liver to form hippurate, and the rapid clearance of hippurate (nitrogen) helps to restore homeostasis [5]. Whether mechanisms other than this also contribute to the beneficial effects of benzoate is not clear.

[1] These studies were performed, in part, at Loyola University Medical Center, and at the University of Arkansas for Medical Science.

Recent studies have shown that benzoate and hippurate have a profound effect on the kidney [7]. This suggests that regulation of nitrogen metabolism or elimination at the level of the kidney may also be an important part of the mechanism of the action of hippurate. The purpose of this study was 2-fold: (1) to investigate the interorgan flow of glutamine, glutamate, urea, and glutathione across the liver, gut, and kidney, and (2) to examine the hypothesis that the interorgan flow of these metabolites can be redirected by harnessing the excretory and metabolic capacity of the kidney, with an emphasis on the role of the γ-glutamyl cycle.

## Methods

Experiments were performed on male Sprague-Dawley rats (280–320 g) [1, 2]. Balance studies of metabolite fluxes across the liver, kidney, and gut were performed by simultaneous measurement of blood flows and arteriovenous concentration differences, in animals that were in the postabsorptive state (12 h fasted) and under inactin anesthesia. The surgical technique employed in the current study has been described previously [1–3]. One group of animals were used to monitor metabolite fluxes across the kidney, and a second group was used to monitor fluxes across liver. Blood (1 ml) was collected in iced heparinized syringes as previously described, and treated with ice-cold TCA [1, 2]. Samples across the kidney were collected at 30-min intervals. One group of animals were infused with benzoate (B), a second group with hippurate (H), and a third group with mannitol (M). In all cases the infusion solution was at a pH of 7.4, and delivered at a rate of 5 μmol/min/100 g b.w. Three samples were collected prior to infusion (which served as baseline control levels) and three samples were collected during infusion of B, H, or M. Measured values for the left kidney were multiplied by 2 to calculate the total values for both kidneys. All results for the kidney represent the mean of two control or experimental periods per animal for B, H or M treatments. Net values represent differences between inflow (IN) and outflow (OUT) and are in nmol/min/100 g b.w. Hypotheses that the mean values for the net output were not significantly different from zero were tested using the t test. Rat renal proximal tubules were prepared exactly as described previously [12]. The results presented in table 4 have been published previously [7].

## Results and Discussion

Table 1 shows the CA, PV, and HV concentration of glutamate, glutamine, urea, and glutathione [1, 2] in plasma. Consistent with previous observations, glutamine was taken up by the liver, gut, and kidney [3]. In contrast, glutamate was released by the liver and taken up by the gut and kidney. The results showed that a significant net synthesis of urea took place only in the liver. As previously described, GSH was released by the liver and gut into

*Table 1.* Hepatic, gut, and renal handling of glutamate, glutamine, urea, and glutathione: net substrate extraction or product release in plasma

| | Liver | | | | | | Gut | Kidney |
|---|---|---|---|---|---|---|---|---|
| | CA | PV | HV | IN | OUT | net | net | net |
| Glutamate (n = 3) | 118 ±7 | 108 ±6 | 181 ±5 | 484 ±29 | 777 * ±23 | 292 ±20 | –36 ±4 | –69 ±6 |
| Glutamine (n = 4) | 378 ±42 | 336 ±25 | 269 ±15 | 1,525 ±134 | 1,157 * ±63 | –368 ±79 | –88 ±41 | –179 ±8 |
| Urea (n = 4) | 7,098 ±517 | 5,990 ±788 | 6,853 ±556 | 27,774 2,565 | 29,658 ±2,299 | 1,885 ±410 | –973 ±1,620 | –104 ±117 |
| Glutathione (n = 7) | 6.7 ±0.2 | 7.1 ±0.5 | 18.4 ±0.6 | 30.0 ±1.6 | 72.3 * ±5.2 | 42.4 ±5.4 | 4.4 ±1.0 | –11.5 ±0.9 |

Results are means ± SEM. Concentrations are in μ*M*; IN and OUT rates are in nmol/min/100 g b.w. from normal rats. Minus sign indicates uptake. * $p < 0.05$ with reference to inflow (IN). CA = Carotid artery; PV = portal vein; HV = hepatic vein.

*Table 2.* Effect of B, H or M on whole-blood arterial, urine acid-base parameters and renal hemodynamics

| Rat group | Blood | | | Urine | | | | | |
|---|---|---|---|---|---|---|---|---|---|
| | pH unit | $HCO_3^-$ m*M* | RBF ml/min | pH unit | $HCO_3^-$ nmol/ min/ 100g | $NH_4^+$ nmol/ min/ 100g | urea nmol/ min/ 100g | GFR ml/min | FLOW μl/min |
| C (n = 12) | 7.41 ±0.01 | 22.6 ±0.7 | 11.6 ±0.3 | 5.9 ±0.1 | 0.4 ±0.1 | 198 ±12 | 651 ±97 | 1.5 ±0.1 | 3.9 ±0.5 |
| + B (n = 4) | 7.44 ±0.03 | 22.1 ±1.0 | 15.0 * ±1.8 | 6.0 ±0.1 | 2.4 * ±1.4 | 453 * ±23 | 2,428 * ±248 | 3.0 * ±0.4 | 19.2 * ±2.1 |
| + H (n = 4) | 7.47 * ±0.01 | 20.5 ±0.8 | 17.9 * ±1.0 | 6.1 ±0.1 | 3.2 * ±0.9 | 711 * ±28 | 3,593 * ±383 | 3.5 * ±0.2 | 25.3 * ±5.8 |
| + M (n = 2) | 7.44 ±0.01 | 18.8 ±0.2 | 12.1 ±2.5 | 6.3 * ±0.3 | 2.6 * ±0.3 | 435 * ±5 | ND | 1.7 ±0.1 | 19.5 * ±0.5 |

Results are means ± SEM of two successive control (C) periods (infused with saline) and two successive benzoate (B), hippurate (H), or mannitol (M) infusions. Results for ammonia are for whole blood. * $p \leq 0.05$ significantly different from control period.

*Table 3.* Effect of B, H or M on renal arteriovenous substrate concentration differences: net renal substrate extraction or product release

| Rat group | Glutamine | | | Glutamate | | | Ammonia | | | Urea | | |
|---|---|---|---|---|---|---|---|---|---|---|---|---|
| | A | A-V | net | A | A-V | net | A | A-V | net | A | A-V | net |
| C | 424 | 79 | –179 | 121 | 49 | –69 | 157 | 95 | 602 | 6,589 | 118 | –104 |
| (n = 12) | ±12 | ±4 | ±8 | ±5 | ±3 | ±6 | ±9 | ±8 | ±29 | ±177 | ±75 | ±117 |
| + B | 440 | 135 | –445* | 101 | 11 | 47* | 162 | 84 | 897* | 7,457 | 495 | –1,089* |
| (n = 4) | ±18 | ±7 | ±25 | ±15 | ±7 | ±7 | ±19 | ±14 | ±111 | ±937 | ±192 | ±126 |
| + H | 394 | 112 | –354* | 163 | 46 | –140* | 172 | 93 | 1,167* | 7,067 | 464 | –1,644* |
| (n = 4) | ±22 | ±21 | ±32 | ±22 | ±6 | ±17 | ±45 | ±11 | ±99 | ±214 | ±45 | ±136 |
| + M | 435 | 70 | –191 | 188 | 76 | –134* | 124 | 68 | 702 | ND | ND | ND |
| (n = 2) | ±33 | ±7 | ±2 | ±6 | ±11 | ±23 | ±12 | ±23 | ±135 | | | |

Results are means ± SEM of two successive control (C) periods (infused with saline) and two successive benzoate (B), hippurate (H), or mannitol (M) infusions. A = Renal artery; V = renal vein; A-V = arteriovenous concentrations are in μ*M*. Net is in nmol/min/100 g b.w. Minus sign indicates uptake. *$p \leq 0.05$, significantly different from control period. Results for Glu and Gln are sum of amounts extracted from plasma plus excreted; $NH_4$ are whole blood plus excreted, and urea represents the amounts extracted from plasma only.

plasma, and taken up from the plasma by the kidney. When fluxes in whole blood were considered, GSH was released by the liver into blood cells at a rate that was ~50% of that of hepatic urea release [1]. The GSH concentration in plasma was trivial as compared to that of whole blood [1]. The release of GSH into red blood cells may serve to protect this compound from being metabolized by the γ-glutamyltransferase (GGT) localized in blood vessels [8].

Baseline values for normal control (C), B, H or M infused animals are presented in table 2. Blood pH and $HCO_3$ excretion were significantly increased by H. Treatment with B or M was associated with a significant increases in urine pH and/or $HCO_3$ excretion. Benzoate or H also had a significant effect on renal hemodynamics. RBF was increased by 1.3-fold by B and 1.5-fold by H. Glomerular filtration rate (GFR) was increased 2-, 2.3-, and 1.1-fold by B, H, or M, respectively. These three treatments were associated with respective 5-, 6.5-, and 5-fold increases in urinary flow rate. Both B and H augmented the rate of urinary excretion of $NH_4$, $HCO_3$, and urea. An enhancement of the urine flow rate by H has previously been described in isolated perfused rabbit tubules [9]. The effects of M on the urinary excretion of $NH_4$ and

*Table 4.* Effect of B, H and AT-125 on the utilization of glutamine (2 m*M*) and the production of glutamate and ammonia in purified preparation of rat renal proximal tubules

| Experimental conditions | Metabolite removal or production nmol/mg protein/30 min | | | |
|---|---|---|---|---|
| | glutamine | glutamate | ammonia | hippurate |
| Glutamine | –309 ± 9 | 127 ± 6 | 363 ± 15 | 0.17 ± 0.1 |
| Glutamine + benzoate | –393 ± 10* | 184 ± 3* | 550 ± 22* | 1.0 ± 0.1* |
| Glutamine + benzoate + AT-125 (0.5 m*M*) | –176 ± 13* | 57 ± 10* | 289 ± 15* | ND |
| Glutamine + hippurate (1 m*M*) | –430 ± 10* | 215 ± 10* | 430 ± 9* | ND |
| Glutamine + AT-125 (0.5 m*M*) | –184 ± 9* | 44 ± 0.1* | 270 ± 14* | ND |

Results are means ± SEM. The tubules were incubated in the presence and absence of modulators (hippurate or benzoate), or inhibitors (AT-125), and the metabolism of glutamine by GGT investigated. Minus sign indicates utilization. Hippurate concentration is in m*M*. * Significantly different from control, $p \leq 0.05$, n = 4.

$HCO_3$ and on urine flow were of a similar magnitude to those produced by the infusion of an equimolar concentration of B, thus suggesting that osmotic effects on renal hemodynamics may partly account for the effects of B and H on nitrogen excretion.

Table 3 shows the arteriovenous plasma concentration differences across the kidney for the major nitrogen-carrying metabolites glutamine, glutamate, ammonia, and urea. Both B and H had significant effects on the renal handling of these metabolites. In the presence of benzoate the kidney becomes a net glutamate-releasing site. This is because B, but not H, inhibits gluconeogenesis [13]. A similar effect has been reported for the isolated perfused rat liver [10]. Equimolar concentrations of M also increased renal glutamate uptake. B and H augmented glutamine extraction by 2.5- and 2-fold, respectively, and ammonia production by 1.5- and 1.9-fold, respectively. The urine-to-plasma ratio for $NH_4$ increased from 0.5 for C to 1.6 for H, indicating a luminal effect of H. The urea extraction rate was increased by 11- and 16-fold by B and H, respectively.

The mechanism of enhanced glutamine extraction or ammonia production was further investigated in rat renal proximal tubules [7, 12]. Proximal tubules exposed to benzoate exhibited an increase in glutamine utilization, and a

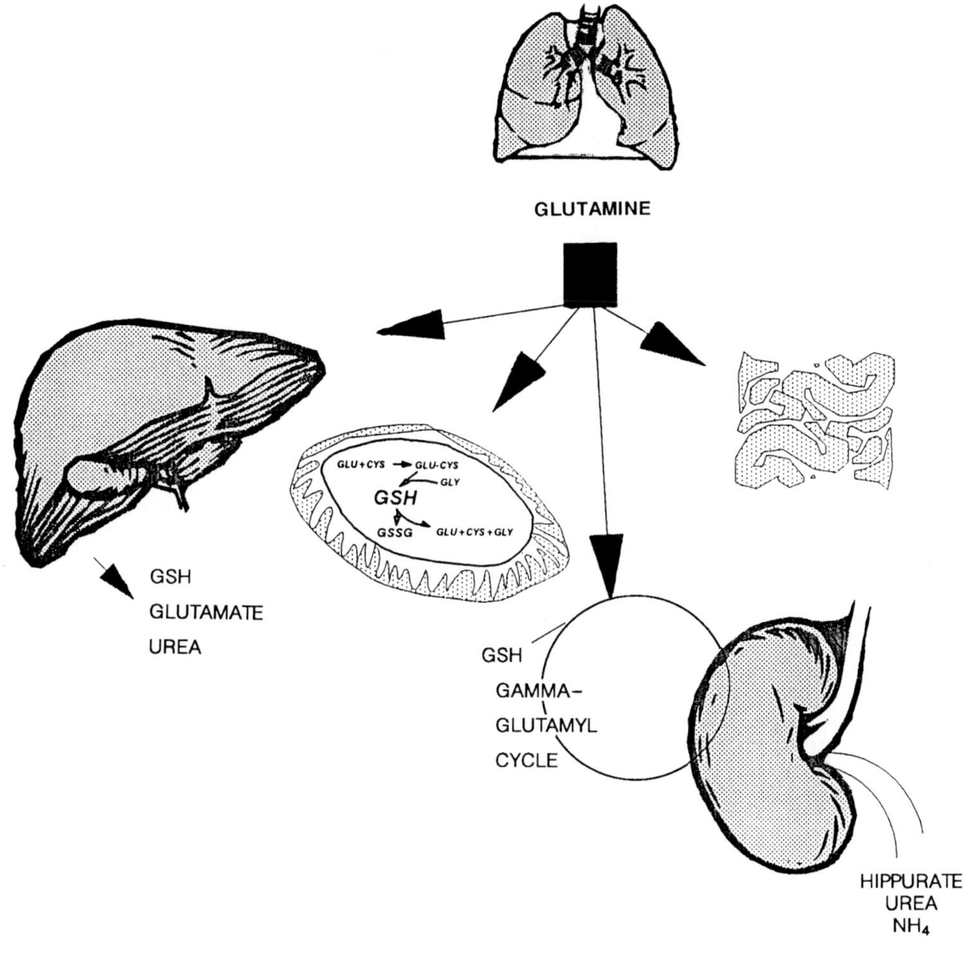


*Fig. 1.* Effect of hippurate on the regulation of interorgan nitrogen flow.

corresponding increase in glutamate and $NH_4$ production (table 4). The mechanism by which B and/or H augmented glutamine-dependent ammoniagenesis, and glutamate production was investigated by using a specific irreversible inhibitor of GGT, acivicin (AT-125) [4]. It is notable that AT-125 is reported to have no effect on the phosphate-dependent glutaminase, a second pathway for ammonia production from glutamine. AT-125 reduced the stimulatory effect of B on glutamine utilization, glutamate and ammonia production. Similar changes were also observed in tubules incubated with glutamine and AT-125 alone, thus indicating that basal levels of glutamine uptake are also reduced by the GGT inhibitor. Since B does not directly affect GGT the effect of B on H synthesis was investigated. The addition of benzoate to tubules resulted in H formation as shown by a ~6-fold increase in H concentration [7]. The synthesis of H from B has been previously reported in kidney slices [11]. When 1 m*M* H (i.e., a concentration similar to that which would be expected to arise from the synthesis of H from B during a 30-min incubation) was added along with glutamine, a stimulatory affect of H on glutamine utilization was observed. Glutamate and ammonia production were increased to levels comparable to those produced by B.

This study supports the hypothesis that the interorgan flow of several important nitrogen-containing metabolites can be redirected by agents that effect the excretory and metabolic capability of the kidney. Figure 1 depicts the interorgan flow of the major nitrogen-carrying metabolites urea and glutamine. H and B augment the excretory capacity of the kidney by increasing renal blood flow and GFR and also increase the metabolic capacity of the kidney by activating the glutaminase activity of GGT, a key enzyme of the modified γ-glutamyl cycle [4]. Together, these renal mechanisms act to enhance renal nitrogen excretion. Further details concerning the interorgan flow of nitrogen metabolites are presented in references 1–4.

In conclusion, these studies establish the following points: (1) the enhanced urinary excretion of urea and $NH_4$ by B treatment is due, in part, to the actions of H; (2) B and H augment $NH_4$ excretion by activation of the glutaminase activity of GGT [14, 15]; (3) H, which appears to be more effective than B in promoting renal nitrogen excretion, and appears to be less toxic [13], may be a useful alternative as a therapeutic agent, and (4) B or H also augment urea excretion by increasing urine output and RBF.

## References


1 Dass PD, Bermes EW, Holmes EW: Renal and hepatic output of glutathione in plasma and whole blood. Biochim Biophys Acta 1992;1156:99–102.

2 Dass PD, Holmes EW, Bermes EW: Hepatic and renal regulation of metabolite flow in the remnant kidney model of chronic renal failure. Contrib Nephrol. Basel, Karger, 1991, vol 92, pp 93–102.

3 Welbourne TC: Interorgan glutamine flow in metabolic acidosis. Am J Physiol 1987;253: F1069–F1076.

4 Dass PD, Martin D: Adaptive ammoniagenesis in chronic renal failure. Renal Physiol Biochem 1990;13:259–263.

5 Batshaw ML, et al: Treatment of inborn errors of urea synthesis. N Engl J Med 1982;306: 1387–1392.
6 Wolff JA, Kulovich S, Yu AL, Qiao CN, Nyhan WL: The effectiveness of benzoate in the management of seizures in nonketotic hyperglycinemia. Am J Dis Child 1986;140:596–602.
7 Dass PD, Bautista NS, Hardman SH, Lawson LR, Kurtz I: Benzoate modulates renal and extrarenal nitrogen flow: Metabolic mechanisms. Life Sci 1991;48:723–731.
8 Dass PD, Misra RP, Welbourne TC: Presence of γ-glutamyltransferase in the renal microvascular compartment. Can J Biochem 1981;59:383–386.
9 Grantham JJ: Fluid secretion in the nephron: Relation to renal failure. Curr Top Physiol 1976;56:248–258.
10 Haussinger D, Stehle T, Colombo J-P: Benzoate stimulates glutamate release from perfused rat liver. Biochem J 1989;264:837–843.
11 Spustova V, Gerykova M, Dzurik R: Hippurate synthesis by rat kidney cortex slices. Molecular Nephrology. Berlin, de Gruyter, 1987, pp 207–212.
12 Dass PD, Lawson LR, Delaney V, Bourke E: Effects of inhibition and modulation of γ-glutamyltransferase on glutamine and glutamate metabolism in control and acidotic rat proximal tubules. Miner Electrolyte Metab 1987;13:433–441.
13 McCune SA, Durant PJ, Flanders LE, Harris RA: Inhibition of hepatic gluconeogenesis and lipogenesis by benzoic acid, *p*-tertbutylbenzoic acid, and a structurally related hypolipidemic agent SC-33459. Arch Biochem Biophys 1982;214:124–133.
14 Silbernagl S: Ammoniagenesis catalyzed by hippurate activated by γ-glutamyltransferase in the lumen of the proximal tubule. Pflugers Arch 1986;407:S72–S79.
15 Welbourne TC, Dass PD: Gamma glutamyltransferase contribution to renal ammoniagenesis in vivo. Pflugers Arch 1988;411:573–578.

P. D. Dass, Miles Inc., Agriculture Division, Toxicology Department,
Stilwell, KS 66085 (USA)

Tizianello A, Baverel G, Endou H, Schoolwerth AC, O'Donovan DJ (eds):
Renal Ammoniagenesis and Interorgan Cooperation in Acid-Base Homeostasis.
Contrib Nephrol. Basel, Karger, 1994, vol 110, pp 158–169

# Organ-Specific Differences in the pH Control of Liver and Kidney α-Ketoglutarate Dehydrogenase

*Brenda C. Smith*[a], *David L. Hyatt*[b], *Joseph Y. Cheung*[b,c], *K.F. LaNoue*[b]

[a] Division of Nephrology, Allegheny General Hospital, Pittsburgh, Pa., and Departments of [c] Medicine and [b] Cellular and Molecular Physiology, The Milton S. Hershey Medical Center, The Pennsylvania State University, Hershey, Pa., USA

Numerous studies have shown that physiological levels of $Ca^{2+}$ in the cell cytosol stimulate pyruvate dehydrogenase, α-ketoglutarate dehydrogenase (α-KGDH) and isocitrate dehydrogenase [1–3]. When cytosolic free $Ca^{2+}$ levels increase, these increases are mirrored inside mitochondrial [4–6] and therefore, under appropriate circumstances, increases in cell $Ca^{2+}$ can stimulate flux through the citric acid cycle [4, 7]. Because of the particularly profound influence of $Ca^{2+}$ on α-KGDH [4] and because tissue levels of α-ketoglutarate (α-KG) may inhibit glutamate dehydrogenase [8] and aspartate aminotransferase [9], it appears that certain pathways of hepatic intermediary metabolism other than the citric acid cycle, such as gluconeogenesis, ureagenesis and the malate-aspartate cycle, are also subject to control by $Ca^{2+}$ [10–12]. $Ca^{2+}$ influences the activity of α-KGDH by lowering the $K_m$ for α-KG [13]. In this way, flux through the citric acid cycle may continue as needed for ATP production while the varying level of α-KG may influence other pathways such as gluconeogenesis. It is perhaps less widely appreciated that $H^+$ ions as well as $Ca^{2+}$ ions influence the $K_m$ of α-KG [13–15]. Just as the increase in $Ca^{2+}$ can induce an increase in gluconeogenesis in the liver, via feedback interactions between α-KG, aspartate aminotransferase, and glutamate dehydrogenase [12, 16], an increase in $H^+$ can also stimulate gluconeogenesis and ammoniagenesis in the kidney via feedback loops [14, 17, 18] analogous to those observed in the liver.

Hormonally induced increases in cytosolic free $Ca^{2+}$ induce 5-fold decreases in α-KG levels in the liver [11, 12, 19] and these decreases are kinetically competent to account for hormonally induced increases in gluconeogenesis and ureagenesis [11, 12]. In a similar manner, the fall of α-KG, which occurs during acute acidosis in the kidney, may account for observed stimulation in phosphate-dependent glutaminase, glutamate dehydrogenase and therefore the ammonia production needed to reverse whole-body acidosis [14, 17, 18].

Conversely, acidosis inhibits hepatic ureagenesis and gluconeogenesis. The physiological importance of this inhibition, which potentially spares bicarbonate and glutamine for conversion to ammonia in the kidney, has been debated [20, 21]. The inhibition occurs and is somewhat paradoxical since $H^+$ has been shown to stimulate α-KGDH in the liver [13]. It seemed strange to us that a $Ca^{2+}$-induced stimulation of gluconeogenesis would occur in the liver secondary to a decrease in the $K_m$ of α-KGDH, whereas a decrease in pH which should produce a similar α-KGDH $K_m$ decrease actually inhibits gluconeogenesis. Therefore, the studies described in this report and in a previous report [22], explore the organ-specific differences in regulation of α-KGDH comparing isolated liver and kidney mitochondria. The tentative conclusion drawn in previous studies [22] was that differences in the control properties are due to organ-specific differences in the degree of complex formation between α-KGDH and mitochondrial aspartate aminotransferase. This hypothesis is supported by new data presented here.

## Methods

*Preparation of Mitochondria*

The liver or kidneys of male Sprague-Dawley rats were perfused in situ with buffer containing 225 m*M* mannitol, 75 m*M* sucrose, 4 m*M* EGTA and 5 m*M* MOPS pH 7.0 at 4 °C. Liver and kidney cortical mitochondria were prepared using standard techniques of differential centrifugation [23]. Respiratory control ratios (RCRs) were determined [23] and only those mitochondria with RCRs > 5 were used. Protein was measured by a biuret assay [24].

*Measurement of 2-OGDH Flux*

Mitochondria were incubated at 28 °C in a buffer containing 145 m*M* KCl, 10 m*M* Tris, 20 m*M* MOPS, 10 m*M* $KH_2PO_4$, 1 m*M* EGTA, and 2.5 m*M* $MgCl_2$. Substrates consisted of 5 m*M* malate, 1 m*M* [2, 3-$^3$H] glutamate, and 0.5 m*M* [U-$^{14}$C]α-KG. The buffer pH was either 6.6 or 7.0.

Samples were taken at 0, 2, 4, 6, and 8 min and extracted with perchloric acid. The extracts were passed through Dowex-1 formate columns and labelled metabolites separated and $^3$H and $^{14}$C radiolabel determined. Calculations of metabolic fluxes were carried out as described previously [22]. Briefly, to determine the rate of $^3$H-α-KG oxidation, the amounts of $^3$H-succinate and $^3$H-malate at 0, 2, and 4 min were determined. Oxidation of $^{14}$C-α-KG was calculated in a similar manner.

In the absence of amino-oxyacetate, aspartate and α-KG were assayed by standard nicotinamide-nucleotide-linked enzymatic reactions [25, 26]. When aspartate was measured in the presence of amino-oxyacetate, acetaldehyde was included in the assay mixture. Total α-KGDH flux was calculated from changes in α-KG and aspartate levels over an 8-min incubation according to the equation:

$$\text{Flux} = \Delta\ \text{aspartate} - \Delta\ \alpha\text{-ketoglutarate}.$$

*Preparation of Mitochondrial Aspartate Aminotransferase*

Mitochondrial aspartate aminotransferase was prepared from isolated liver mitochondria as described by Barra et al. [27] and adapted by Porter et al. [28]. After isolation, high-performance liquid chromatography revealed a single protein peak with a molecular weight of approximately 90 kD consistent with the molecular weight of aspartate aminotransferase dimer.

*Assay of Aspartate Aminotransferase Activity*

The activity of aspartate aminotransferase was assayed in buffer consisting of 100 m*M* Hepes, 70 m*M* KCl, 10 m*M* MOPS, 0.5 m*M* Tris, 1.5 m*M* EGTA, 5 m*M* $KH_2PO_4$ in the presence of 20 m*M* α-KG, 30 m*M* aspartate, 140 μ*M* NADH, and 5 units of malate dehydrogenase at pH 7.0, 28 °C. In experiments where intramitochondria aspartate aminotransferase activity was measured, baseline activity (activity of extramitochondrial enzyme) was determined prior to the addition of Triton X-100. These baseline values were substracted from total mitochondrial aspartate aminotransferase activity to determine intramitochondrial activity. Enzyme activity was determined from the change in NADH absorption on a Zeiss spectrophotometer.

*Mitochondrial Loading of Aspartate Aminotransferase*

In order to alter the ratio of α-KGDH/aspartate aminotransferase, kidney mitochondria were loaded with liver mitochondrial aspartate aminotransferase as described by Marra et al. [29] and Passarella et al. [30]. Kidney mitochondria (2.5 mg/ml) were incubated in buffer containing 250 m*M* sucrose, 20 m*M* Tris-HCl, 1 m*M* EGTA, and 1 m*M* ATP, pH 7.4, with 80 μg mitochondrial aspartate aminotransferase/ml buffer, at 34 °C for 3 min. The mitochondria were sedimented via centrifugation, washed and resuspended. The degree of loading was determined by measuring intramitochondrial enzyme activity. Controls consisted of mitochondria incubated with commercially available cystolic aspartate aminotransferase which does not cross the mitochondria membrane [31] and which does not produce an increase in mitochondrial transferase activity, as shown in the present study.

*Statistics*

Values shown are means ± SEM. Unless otherwise indicated, statistical analysis was performed by Student's t test for paired and unpaired data.

## Results

Our previous studies have suggested that α-KG levels can provide important feedback inhibition of gluconeogenesis in liver and kidney. α-KG can also inhibit hepatic urea synthesis and renal ammoniagenesis [8, 9, 11, 12]. Therefore, an obvious first step in the investigation of the pH dependence of these

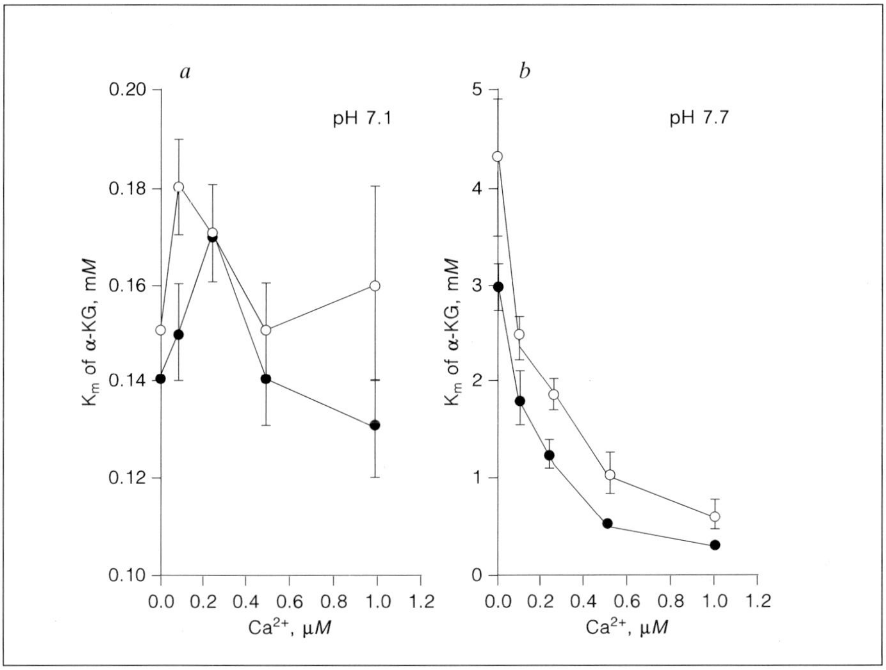


*Fig. 1.* The influence of pH (*a*, pH 7.1; *b*, pH 7.7) and $Ca^{2+}$ on the $K_m$ of rat liver and kidney α-KGDH. The liver data is shown as closed circles and the kidney as open circles. Data is taken from table 1 of Smith et al. [22] and was obtained using Triton X-100 permeabilized mitochondria.

processes was to investigate the cation ($Ca^{2+}$ and $H^+$) dependence of the kidney and liver enzymes themselves in a parallel study. The kinetic properties of the enzymes were investigated in Triton X-100 permeabilized mitochondria so that substrates and cations could be varied independent of permeability barriers. The results are shown in figure 1. Data are adapted from an earlier report of this work [22].

Figure 1 shows the effect of varied $Ca^{2+}$ on the $K_m$ of both enzymes at either pH 7.1 (fig. 1a) or pH 7.7 (fig. 1b). Note that at the lower pH, the $K_m$ is small for both enzymes, and thus the scale of figure 1a is lower than the scale of figure 1b. At the higher pH and at lower $Ca^{2+}$ levels, the $K_m$ for both enzymes is in the m*M* range and drops precipitously to μ*M* levels as $Ca^{2+}$ increases. No significant organ-specific difference could be observed in the influence of cations on the α-KG $K_m$. In contrast to this finding, large differences were found comparing the $V_{max}$s of the enzymes, since the $V_{max}$ of the kidney is about 3–4 times that of the liver. This presumably reflects the fact that there

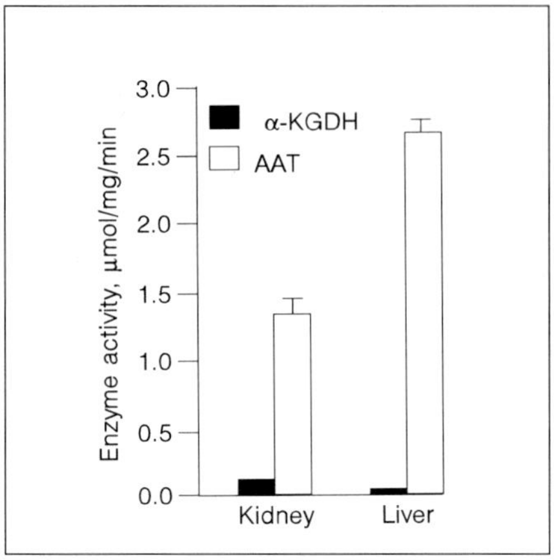


*Fig. 2.* The relative activities of α-KGDH and mitochondrial aspartate aminotransferase in kidney and liver. Activities of aspartate aminotransferase (AAT) and α-KGDH were assayed as described here and in Smith et al. [22].

is a greater amount of the enzyme protein in the kidney. On the other hand, when the $V_{max}$ of mitochondrial aspartate aminotransferase (an important source of α-KG) was measured in both types of mitochondria, it appeared there was less transferase in the kidney. The large organ-specific difference in the ratio of aspartate aminotransferase to α-KGDH is shown in figure 2. The data suggest that the kidney α-KGDH may normally operate far below its $K_m$.

Further experiments were carried out with intact mitochondria to examine the kinetic properties of the enzymes operating in the natural milieu of the kidney or liver mitochondrial matrix. To determine the influence of α-KG transport across the mitochondrial membrane, liver and kidney mitochondria were incubated at 28 °C with low physiological levels of U-$^{14}$C-α-KG (0.5 m*M*) and 2, 3-$^{3}$H glutamate (1 m*M*). The reaction was halted with perchloric acid at 0, 2, 4, 6, and 8 min and extracts subjected to anion exchange chromatography. The disappearance of media $^{14}$C-α-KG and the appearance of $^{3}$H-α-KG provided a minimal estimate of the rate of transport of α-KG across the mitochondrial membrane. The $^{14}$C-α-KG was converted either to $^{14}$C-glutamate or to the products of α-KG oxidation, succinate and malate. Because of the presence of the large unlabelled sink of malate (5 m*M*) in the reaction mixtures, all radiolabelled products were trapped. Corrections were made for loss of $^{14}CO_2$ and $^{3}H_2O$ in the metabolic reaction sequences up to malate.

The effect of pH, $Ca^{2+}$, and ADP on transport rates in liver and kidney mitochondria have been reported previously [22], as have the effect of these parameters on $^{14}$C-glutamate appearance. Measurements of transport provided little insight into the molecular basis for possible organ-specific differences in the metabolism of α-KG. The transport rates were not influenced by pH and $Ca^{2+}$, but were slightly stimulated by ADP in the

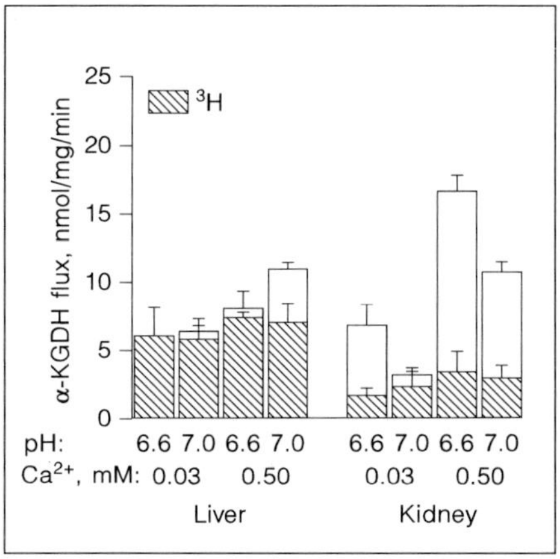


*Fig. 3.* The effect of $Ca^{2+}$ and pH on α-KGDH flux in intact mitochondria. Data were obtained as described in Methods by analyzing the amount of tritium- and $^{14}C$-labelled malate and succinate generated during incubation of either liver or kidney mitochondria with 0.5 m*M* [U-$^{14}$C]-α-KG plus 1 m*M* [2, 3-$^{3}$H]-glutamate. Cross-hatched areas indicate tritium-labelled products.

kidney, suggesting that the $V_{max}$ of the α-KG transporter may have been somewhat higher in the kidney mitochondria than liver mitochondria. The $^{3}H/^{14}C$ ratios of the products of α-KG oxidation were more revealing, as shown in figure 3.

An increase of mitochondrial matrix $Ca^{2+}$ from 0.05 to 0.5 μ*M* $Ca^{2+}$ stimulates the oxidation of α-KG in both types of mitochondria. This is illustrated by the increase in the sum of the $^{3}H$- and $^{14}C$-labelled products. However, the effect of $Ca^{2+}$ is more dramatic in kidney mitochondria than liver. As expected, an increase in media $H^{+}$ from pH 7.0 to 6.6 also stimulates the oxidation of α-KG in kidney. In the case of liver mitochondria, $H^{+}$ did not stimulate α-KG oxidation, but in fact inhibited it. These observations were confirmed by estimates of the overall oxidation of α-KG, as measured by the sum of the appearance of aspartate and disappearance of total α-KG (cf. Methods). The vast majority of the products of α-KG oxidation in the liver are labelled with tritium, indicating that they are derived from intramitochondrial transamination of glutamate, despite the fact that the rate of entry of $^{14}C$-α-KG into the mitochondrial matrix was similar to the rate of generation of $^{3}H$-α-KG, estimated from the rate of production of aspartic acid.

On the other hand, in the kidney mitochondria, the majority of the oxidation products were $^{14}C$ labelled and the $^{14}C/^{3}H$ ratio of the products correlated well with estimates of relative rates of appearance of $^{14}C$-α-KG and $^{3}H$-α-KG. These observations led to the hypothesis that carbon may be directly channeled from glutamate to the products of α-KG oxidation via an enzyme complex composed of aspartate aminotransferase and α-KGDH. Evidence for such a complex has been found previously by Fahien and co-workers [32, 33] working with liver enzymes. The large amount of aspartate aminotransferase

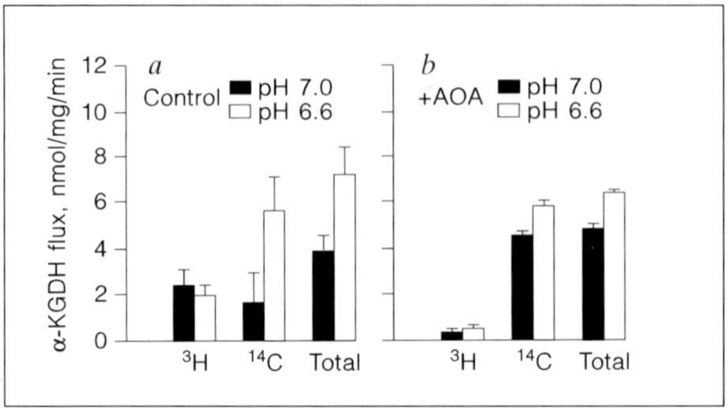


*Fig. 4. a, b* The effect of amino-oxyacetic acid on the pH sensitivity of α-KGDH product formation in kidney mitochondria. Data were obtained as described in Methods.

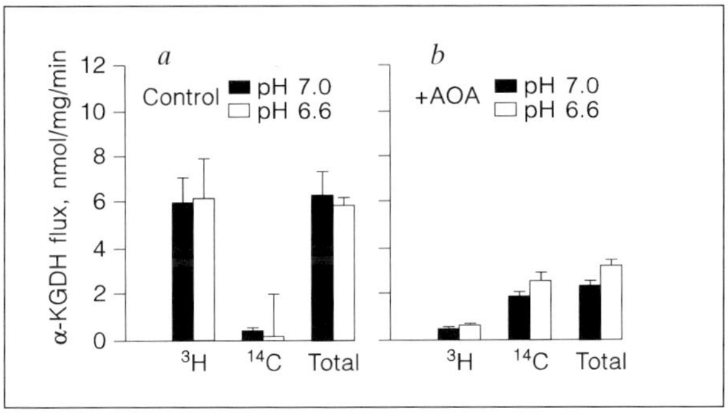


*Fig. 5.a,b* The effect of amino-oxyacetic acid on the pH sensitivity of α-KGDH product formation in liver mitochondria. Data were obtained as described in Methods.

relative to α-KGDH found in liver mitochondria (as opposed to kidney) may favor enzyme complexing in the liver (cf. fig. 2).

Since the glutamate aspartate transporter has an alkaline pH optimum [34], the production of aspartate and therefore $^3$H-labelled α-KG is more rapid at the more alkaline pH in both kidney and liver mitochondria [22]. This and the presumed formation of the complex between the dehydrogenase and the aminotransferase in liver may be the basis for the alkaline pH optimum of α-KG oxidation in the liver. To test this hypothesis, the effect of pH on oxidation of α-KG was measured in the presence of amino-oxyacetic acid, a potent inhibitor of aspartate aminotransferase which penetrates the mitochondrial inner membrane [35]. Control experiments were carried out with *D,L*-cycloserine present in the external media. Cycloserine does not penetrate the mitochondrial membrane but does inhibit transaminase bound to the mitochondria on the outside [36]. The results of the experiment with kidney are shown in figure 4, and for liver are shown in figure 5. The protocol was similar to that described for figure 3, except that the influence of $Ca^{2+}$ was not determined, but the effect of amino-oxyacetate was.

Anion exchange chromatography permitted us to quantitate the amount of $^{3}H$ and $^{14}C$ malate and succinate, as shown in figures 4 and 5. The control data for both types of mitochondria obtained in the presence of cycloserine are not significantly different from the results shown in figure 3. In the presence of amino-oxyacetic acid, aspartate formation was almost completely blocked and in accord with this observation, the majority of the products of α-KG oxidation in both liver and kidney were labelled with $^{14}C$, not $^{3}H$. Most importantly, in the presence of amino-oxyacetic acid, the oxidation of α-KG was stimulated by $H^{+}$ not only in the kidney (fig. 4), but also in the liver (fig. 5).

In order to further test the hypothesis that the pH sensitivity of α-KG oxidation is dependent on the intramitochondrial ratio of aspartate aminotransferase to α-KGDH activities, attempts were made to increase that ratio in the kidney. Rather than lower dehydrogenase activity, efforts were made to increase the amount of aminotransferase in the kidney mitochondria. Previous reports indicate that isolated rat liver mitochondrial aspartate aminotransferase will enter liver mitochondria from the surrounding medium [30]. Apparently the functional sequence necessary for translocation across the mitochondrial membranes is contained within the amino sequences of the mature enzyme. Therefore in an effort to increase the transferase activity in kidney mitochondria, the liver enzyme was isolated [27, 28] and incubated (80 μg/ml) with kidney mitochondria as described in Methods. In contrast to the previous reports, we found it was necessary to include 1 m*M* ATP in the incubation to get significant increases in the kidney matrix aminotransferase activity. Longer incubation times and higher medium enzyme activity did not increase the amount of activity transferred into the kidney mitochondria. About half the attempts failed but significant increases, averaging $21.8 \pm 3.8\%$, were obtained in three trials and the aspartate aminotransferase loaded kidney mitochondria were used to measure the pH sensitivity of α-KGDH. These aminotransferase-loaded mitochondria were compared directly with kidney mitochondria which had been treated in an identical manner, but incubated with the commercially available cytosolic isozyme of aspartate aminotransferase. These incubations resulted in no increase in intramitochondrial transferase activity. As shown in figure 6, the behavior of kidney mitochondria incubated with cytosolic aspartate aminotransferase differed very little from that of the control kidney mitochondria. The kidney mitochondria incubated with liver mitochondrial aspartate aminotransferase on the other hand exhibited altered pH sensitivity. The

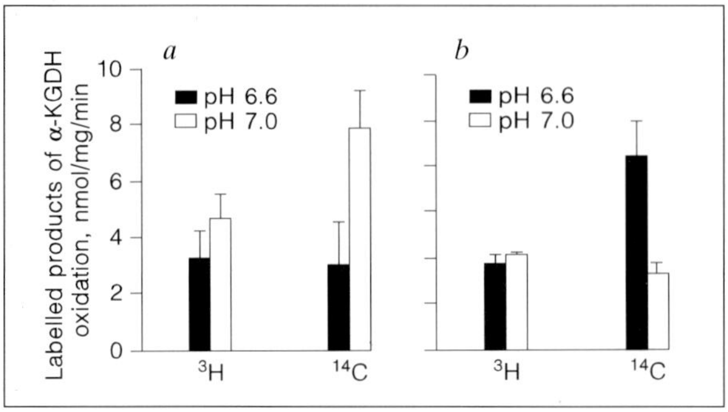


*Fig. 6. a, b* The influence of loading extra (21.8%) liver mitochondrial aspartate aminotransferase into kidney mitochondria on α-KGDH product formation. Data were obtained as described in Methods. *a* = Control; *b* = + AAT

oxidation of α-KG was not stimulated by $H^+$, but was inhibited. The increase in intramitochondrial transaminase activity, however, did not result in an increase in the production of aspartate nor an increase in the amount of $^3H$ products of α-KG oxidation. This is probably reasonable since the rate of glutamate transamination is limited by the entry of $^3H$-glutamate into the mitochondria not by the amount of aspartate aminotransferase inside mitochondria. Surprisingly, there are more $^{14}C$-oxidation products at the more alkaline pH. The $^{14}C$-α-KG in the mitochondria may not have easy access to the α-KGDH complexed with aminotransferase until the ketoacid is converted to $^{14}C$-glutamate inside the matrix. It is also surprising that such a relatively small (22%) increase in aspartate aminotransferase could have caused the observed changes in the pH sensitivity of the dehydrogenase. Perhaps the liver enzyme used to load the kidney mitochondria is more prone to form complexes with the α-KGDH. Clearly further experiments are required to clarify the behavior of the aspartate aminotransferase-loaded kidney mitochondria.

## Discussion

The cells of the kidney and liver share many of the same enzymes in the pathways of gluconeogenesis and ammoniagenesis, but in order to restore balance during periods of acute acidosis, renal gluconeogenesis and ammoniagenesis are stimulated while hepatic gluconeogenesis and ureagenesis are inhibited [14, 18, 20]. On the other hand, increasing $Ca^{2+}$ levels stimulates glucose formation in both organs [37]. It appears likely that $Ca^{2+}$ stimulation of α-KGDH is a key factor in this signalling mechanism. Chronic as opposed to acute acidosis in the kidney induces increased synthesis of key enzymes required in the ammoniagenesis pathway, e.g., glutaminase and phosphoenolpyruvate carboxykinase [38]. Clearly the liver does not respond in this way to acidosis. The studies described here represent an attempt to understand the basis for the acute response of the two organs to acidosis. The focus of the study is on the pH control of α-KGDH since previous studies of ours and others suggest that levels of α-KG which are controlled by $Ca^{2+}$ and $H^+$ may control glutamate dehydrogenase, aspartate aminotransferase and thereby the gluconeogenic and ammoniagenic pathways in both the liver and kidney [10–12, 17, 18]. Our previous studies [22] likewise suggested that the enzyme is activated by $Ca^{2+}$ in both organs but by $H^+$ only in the kidney.

Studies reported here and elsewhere [22] suggested that complex formation between α-KGDH and mitochondrial aspartate aminotransferase and consequent channeling of carbon from the transferase to the dehydrogenase in the liver gives the hepatic α-KGDH the pH sensitivity of the aspartate aminotransferase rather than its own pH sensitivity. Tests of the hypothesis described here support this concept. Decreasing aspartate aminotransferase activity in the liver with amino-oxyacetic acid altered the pH sensitivity of α-KG oxidation in intact liver mitochondria to resemble that in kidney. On the other hand, increasing mitochondrial aspartate aminotransferase in the kidney mitochondria alters the pH sensitivity of kidney so that it resembles that of the liver. The success of the transferase loading experiment was surprising however in view of the fact that only a 22% increase in transferase activity caused a complete reversal of the pH response. We expected at least a doubling of the transferase activity would be required, assuming that only the relative amounts of the two enzymes determine complex formation.

Fahien and co-workers [32, 33] have studied complex formation between these two liver enzymes as well as other liver mitochondrial enzymes. The method used by Fahien and co-workers [32, 33] to detect complex formation is to solubilize one enzyme, e.g., aspartate aminotransferase, in polypropylene solutions, after which various amounts of a second enzyme are added. The degree of complex formation is assessed by precipitation of the relevant activities from solution as insoluble complexes. From these studies one would predict that, even in the kidney, complex formation between mitochondrial aspartate aminotransferase and α-KGDH would involve all the α-KGDH. Of course, the environment of the mitochondrial matrix is far different from that of the polypropylene solution. Nevertheless it is possible that the liver transferase differs somehow from the kidney transferase, making it more liable to form α-KGDH complexes. Another possibility is that complex formation between entering mitochondrial aspartate aminotransferase and the membrane-bound α-KGDH is sterically favored.

Possibly these questions can be more definitively answered with molecular biological techniques using appropriate cell lines in which one could induce excess mitochondrial aspartate aminotransferase gene expression. Further studies using improved techniques are required.

## References


1 Hansford RG: Relation between mitochondrial calcium transport and control of energy metabolism. Rev Physiol Biochem Pharmacol 1985;102:1–72.

2 Gunter TE, Pfeiffer DR: Mechanisms by which mitochondria transport calcium. Am J Physiol 1990;258:C755–C786.

3 Denton RM, McCormack JG: $Ca^{2+}$ transport by mammalian mitochondria and its role in hormone action. Am J Physiol 1985;249:E543–E554.

4 Wan B, LaNoue KF, Cheung JY, Scaduto RC: Regulation of citric acid cycle by calcium. J Biol Chem 1989;264:13430–13439.

5 Lukacs GL, Kapus A: Measurement of the matrix-free $Ca^{2+}$ concentration in heart mitochondria by entrapped fura-2 and quin-2. Biochem J 1987;248:609–613.

6 Peers M, Kelly RA, Smith TW: Calcium and proton activities in rat cardiac mitochondria. Effect of matrix environment on behavior of fluorescent probes. Biochem J 1989;257:131–142.

7 Moreno-Sanchez R, Hansford RG: Dependence of cardiac mitochondrial pyruvate dehydrogenase activity on intramitochondrial free $Ca^{2+}$ concentration. Biochem J 1988;256:403–412.

8 Schoolwerth AC, LaNoue KF: Control of ammoniagenesis by α-ketoglutarate in rat kidney mitochondria. Am J Physiol 1983;244:F399–F408.

9 Strzelecki T, Strezelecka D, Koch CD, LaNoue KF: Sites of action of $Ca^{2+}$ mobilizing hormones on gluconeogenesis. Arch Biochem Biophys 1988;264:310–320.

10 Kneer NM, Wagner MJ, Lardy HA: Regulation by calcium of hormonal effects on gluconeogenesis. J Biol Chem 1979;254:12160–12168.

11 Sterniczuk A, Hreniuk S, Scaduto RC, LaNoue KF: The mechanism of $Ca^{2+}$-related control of gluconeogenesis in perfused rat liver. Eur J Biochem 1991;196:143–150.

12 Sterniczuk A, Hreniuk S, Scaduto RC, LaNoue KF: Effect of phenylephrine on pyruvate dehydrogenase in fasting rat livers. Eur J Biochem 1991;196:151–157.

13 McCormack JG: Characterization of the effects of $Ca^{2+}$ on the intramitochondrial $Ca^{2+}$-sensitive enzymes from rat liver and within intact rat liver mitochondria. Biochem J 1985;231:581–595.

14 Lowry M, Ross BD: Activation of oxoglutarate dehydrogenase in the kidney in response to acute acidosis. Biochem J 1980;190:771–780.

15 Rutter GA, Denton RM: Regulation of $NAD^+$-linked isocitrate dehydrogenase and 2-oxoglutarate dehydrogenase by $Ca^{2+}$ ions within toluene-permeabilized rat heart mitochondria. Interactions with regulation by adenine nucleotides and $NADH/NAD^+$ ratios. Biochem J 1988;252:181–189.

16 Hreniuk S, LaNoue KF: Feedback control of pyruvate carboxylase by oxalacetate. International Congress of Biochemistry, Prague 1988, abstr MO-95:82.

17 Preuss HG, Vavatsi-Manos O, Vertuno CL, Baird K: The effects of pH change on renal ammoniagenesis in vitro. Proc Soc Exp Biol Med 1974;146:803–808.

18 Tannen RL, Kunin AS: Effect of pH on ammonia production by renal mitochondria. Am J Physiol 1976;231:1631–1637.

19 Taylor WM, Reinhart PH, Bygrave FL: Stimulation by α-adrenergic agonists of $Ca^{2+}$ fluxes, mitochondrial oxidation and gluconeogenesis in perfused rat liver. Biochem J 1983;212:555–565.

20 Häussinger D, Gerok W, Sies H: Hepatic role in pH regulation: Role of the intracellular glutamine cycle. Trends Biochem Sci 1984;9:300–302.
21 Halperin ML, Jungas RL, Cheema-Dhadli S, Brosnan JJ: Disposal of the daily acid load: An integrated function of the liver, lungs and kidney. Trends Biochem Sci 1987;12:197–199.
22 Smith BC, Clotfelter LA, Cheung JY, LaNoue KF: Differences in 2-oxoglutarate dehydrogenase regulation in liver and kidney. Biochem J 1992:284:819–826.
23 Schoolwerth AC, Nazar BL, LaNoue KF: Glutamate dehydrogenase activation and ammonia formation by rat kidney mitochondria. J Biol Chem 1978;253:6177–6183.
24 Cleland KW, Slater EC: Respiratory granules of heart muscle. Biochem J 1953;53:547–556.
25 Williamson JR, Corkey BE: Assays of intermediates of the citric acid cycle and related compounds by fluorometric enzyme methods. Methods Enzymol 1969;13:434–513.
26 Lowry OH, Passoneau JV: A collection of metabolite assays; in A Flexible System of Enzymatic Analysis. New York, Academic Press, 1972, pp 146–218.
27 Barra D, Bossa F, Doonan S, Fahmy HM, Martini F, Hughes GJ: Large-scale purification and some properties of the mitochondrial aspartate aminotransferase from pig heart. Eur J Biochem 1976;64:519–526.
28 Porter PB, Barra D, Bossa F, Cantalupo G, Doonan S, Martini F, Sheehan D, Wilkinson SM: Purification and basic properties of the aspartate aminotransferase from a variety of sources. Comp Biochem Physiol 1981;69B:737–746.
29 Marra E, Doonan S, Saccone C, Quagliariello E: Studies of the selective permeation of radioactively labelled aspartate aminotransferase isozymes into mitochondria in vitro. Eur J Biochem 1978;83:427–435.
30 Passarella S, Marra E, Atlante A, Barile M, Doonan S, Quagliariello E: Uptake of aspartate aminotransferase into mitochondria in vitro causes efflux of malate dehydrogenase and vice versa. Biochim Biophys Acta 1990;1022:273–282.
31 Marra E, Doonan S, Saccone C, Quagliariello E: Selective permeability of rat liver mitochondria to purified aspartate aminotransferases in vitro. Biochem J 1977;164:685–691.
32 Fahien LA, Kmiotek EH, McDonald MJ, Fibich B, Mandic M: Regulation of malate dehydrogenase activity by glutamate, citrate, α-ketoglutarate and multienzyme interaction. J Biol Chem 1988;263:10687–10697.
33 Fahien LA, Teller JK: Glutamate-malate metabolism in liver mitochondria. A model constructed on the basis of mitochondrial levels of enzymes, specificity, dissociation constants, and stoichiometry of hetero-enzyme complexes. J Biol Chem 1992;267:10411–10422.
34 Tischler ME, Pachence J, Williamson JR, LaNoue KF: Mechanism of glutamate-aspartate translocation across the mitochondrial inner membrane. Arch Biochem Biophys 1976;173: 448–462.
35 Hopper S, Segal HL: Kinetic studies of rat liver glutamic-alanine transaminase. J Biol Chem 1962;237:3189–3195.
36 LaNoue KF, Meijer AJ, Brouwer A: Evidence for electronic aspartate transport in rat liver mitochondria. Arch Biochem Biophys 1974;161:544–550.
37 Kurokawa K, Massry SG: Evidence for stimulation of renal gluconeogenesis by catecholamines. J Clin Invest 1973;52:961–964.
38 Kaiser S, Hwang JJ, Curthoys NP: Effect of acidosis on phosphoenol-pyruvatecarboxykinase and glutaminase mRNAs in rat kidney and LLC-PK-$F^+$ cells; in Endou H, Schoolwerth AC, Baverel G, Tizianello A (eds): Molecular Aspects of Ammoniagenesis. Contrib Nephrol. Basel, Karger, 1991, vol 92, pp 206–210.

Kathryn F. LaNoue, PhD, Professor, Department of Cellular and Molecular Physiology,
The Milton S. Hershey Medical Center, The Pennsylvania State University,
PO Box 850, Hershey, PA 17033 (USA)

Tizianello A, Baverel G, Endou H, Schoolwerth AC, O'Donovan DJ (eds):
Renal Ammoniagenesis and Interorgan Cooperation in Acid-Base Homeostasis.
Contrib Nephrol. Basel, Karger, 1994, vol 110, pp 170–175

# The Hepatic and Renal Response to Hydrochloric Acid Ingestion by the Rat

*Anne L. Lardner, Daniel J. O'Donovan*

Department of Physiology, University College, Galway, Ireland

Nitrogenous metabolism is altered during chronic metabolic acidosis [1–3]. The obvious change is the increased renal extraction of glutamine to provide extra ammonia to buffer the hydrogen ions in the renal tubular lumen [4–6]. Glutamine plays a strategic role in nitrogen metabolism in many tissues [7–10]. It is of interest, therefore, to determine the effect of diverting glutamine to the kidney over a prolonged period.

Ammonia is converted to urea in the periportal region of the liver [11, 12]. This process, of itself, should not influence renal ammonia production, since plasma ammonia contributes insignificantly to urinary ammonia. However, the ammonia, which is converted to glutamine in the perivenous region of the liver, is providing a substrate for renal ammoniagenesis. Hepatic ureagenesis and glutamine synthesis share a common substrate, namely ammonia [13]. Thus an increased renal extraction of glutamine and an acceleration of renal glutamine metabolism in acidosis could have an influence on the disposal of ammonia by the liver [14].

Chronic metabolic acidosis is often induced in rats by replacing the drinking water with a 0.28 *M* ammonium chloride solution. The ammonia is converted to urea in the liver and hydrochloric acid is produced systemically. This should be equivalent to administering hydrochloric acid into the alimentary tract. However, it is also priming the synthesis of urea and glutamine in the liver, which could influence the production of ammonia by the kidney. Thus it was of interest to determine the effect of administering hydrochloric acid to rats and to compare it to ammonium chloride administration.

## Methods

Female Sprague-Dawley rats weighing approximately 250 g were placed in metabolic cages for 7 days. During this adjustment period they were exposed to a 12-hour light-dark cycle at constant room temperature. They were offered powdered chow (Redmills, Kilkenny, Ireland) and given tap water ad libitum. They were then assigned to two groups. The control group was given a mixture of powdered pellets and tap water. The experimental group were given HCl.

A number of preliminary experiments indicated that in order for the animals to receive a continuous supply of HCl, it should be mixed with the food. The model chosen to produce chronic metabolic acidosis was an equal mixture of 0.8 *M* HCl and powdered pellets for 8 days. The food intake of the HCl group was monitored during the preliminary period. They were then continued on this regime during the experimental period and the control group were pair fed an equal amount of food. Thus the acidotic group consumed 9.5 mmol/day of HCl.

Food and water intakes were recorded daily. Urine was collected daily in flasks containing thymol preservatives. At the end of the experimental period, rats were anaesthetized with ether and blood samples were obtained by cardiac puncture. The blood was centrifuged at 3,000 *g* for 15 min and the resulting serum was subjected to $smac_{16}$ for urea determination. Following ether overdose, the kidneys were removed for glutaminase assay.

Serum $smac_{16}$ analysis was carried out using a Technicon multiple analyser. Urinary urea was determined by pretreating the sample with urease and assaying the resulting ammonia by the reductive amination of oxoglutarate to glutamate by NADH and sample ammonia using glutamate dehydrogenase. Renal phosphate-dependent glutaminase (PDG) and phosphate-independent glutaminase (PIG) were assayed in whole homogenates by measuring the amount of $NH_3$ released from 0.1 *M* glutamine with and without the addition of 0.17 *M* phosphate.

## Results

The model described in the Methods section was found to be suitable for providing rats with a constant intake of HCl during an 8-day period. The volume of HCl ingested is similar to that of $NH_4Cl$ added to the drinking water. The animals ingested 9.5 mmol HCl daily. Rats ingested a similar amount of $NH_4Cl$, when offered at 0.28 *M* solution, which is the maximum concentration that can be offered without a reduction in fluid intake. The only obvious deficiency in the HCl model is that the animals consume 50% less food than the control group. In these experiments the control group was pair fed with the experimental group. The water intake was the same for the control and acidotic groups.

Figure 1 shows the serum urea concentration on the 8th day of the experiment. Acidosis had no significant effect on the serum urea concentration.

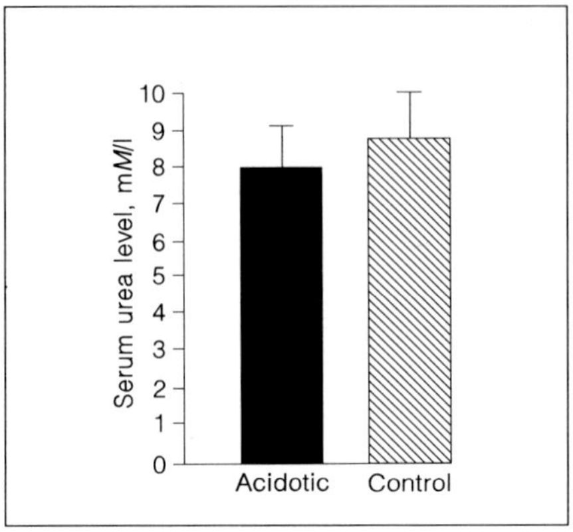


*Fig. 1.* Serum urea concentration (m$M$) of control and acidotic rats. Rats were made acidotic by mixing HCl with the meal. They ingested 9.5 mmol/day for 8 days.

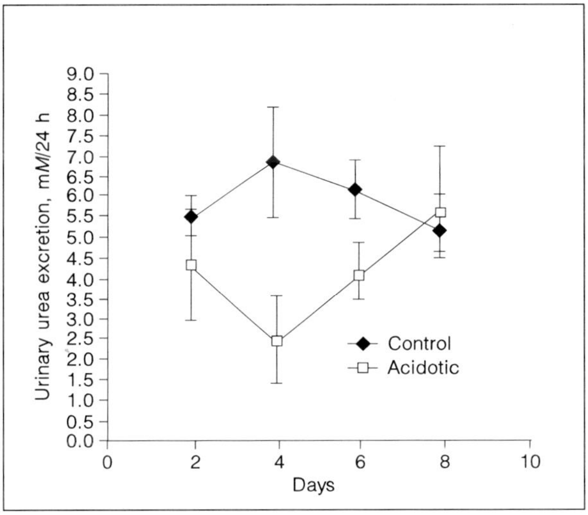


*Fig. 2.* The rate of urinary urea excretion of control and acidotic rats.

However, acidosis decreased the rate of urinary urea excretion for the first 6 days of treatment, which then returned to control value on the 8th day (fig. 2).

Acidosis had no significant effect on the activity of renal PIG (fig. 3). However, the activity of renal PDG was increased significantly by acidosis (fig. 4). Thus this model for HCl acidosis had a similar effect on renal glutaminase as drinking a 0.28 $M$ $NH_4Cl$ solution.

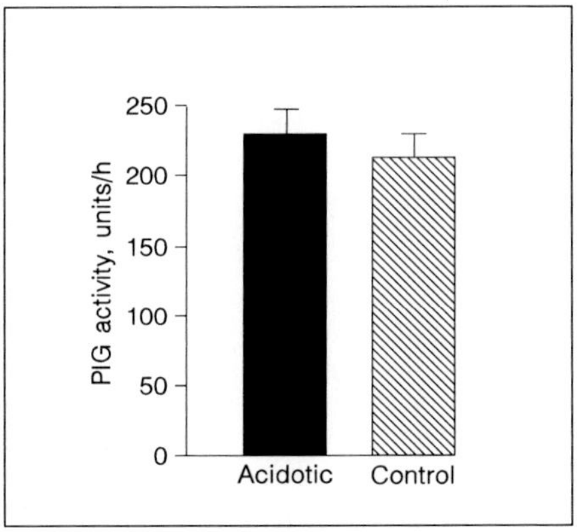


*Fig. 3.* Renal PIG of control and acidotic rats. Glutaminase was assayed by the ammonia released from 0.1 *M* glutamine incubated with kidney homogenates.

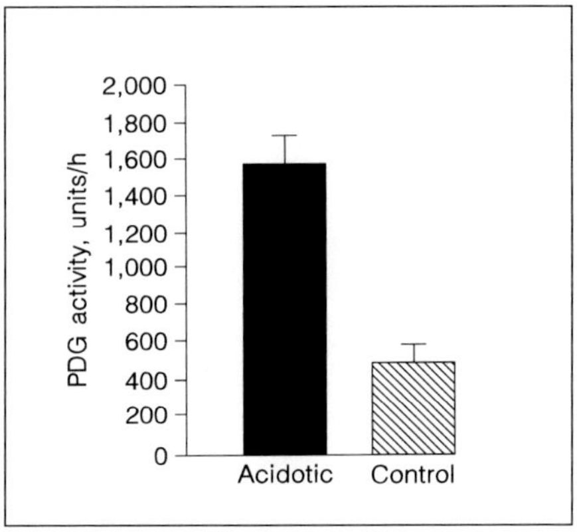


*Fig. 4.* Renal PDG of control and acidotic rats. Incubation conditions differed from figure 3 in that 0.17 *M* phosphate was incubated.

## Discussion

The traditional view that the renal extraction of glutamine and enzymatic adaptation of glutamine metabolism within the renal tubule are the key elements in the control of acidosis has been challenged in recent times [15–17]. Since there is no major change in plasma glutamine concentration in acidosis despite an enhanced renal extraction of this compound raises the possibility of an alteration in extrarenal nitrogen metabolism. Hepatic ammonia is a substrate for both urea and glutamine. Thus the hepatic urea cycle activity may play a part in determining the availability of nitrogenous compounds for renal acid-base regulation. Furthermore, bicarbonate is a substrate for hepatic ureagenesis.

Chronic metabolic acidosis is frequently produced by allowing animals to voluntarily drink $NH_4Cl$ solutions [18, 19]. This procedure, in addition to producing acidosis, provides the substrate, ammonia, for the hepatic synthesis of urea and glutamine. It is conceivable that priming the urea cycle in this way may preclude the effects that changes in this cycle may have on renal ammonia production.

Rats will not normally ingest HCl solutions in sufficient quantities to make them acidotic. Intragastric administration of HCl enters the systemic circulation intermittently [20]. Mixing of HCl with the diet has proven to be a satisfactory means of making rats chronically acidotic.

In these experiments, HCl increased the activity of renal PDG (fig. 4) to the same degree as drinking a 0.28 *M* $NH_4Cl$ solution [21]. Under these conditions acidosis resulted in renal glutaminase adaptation without supplying any ammonia systemically. This procedure decreased the rate of urinary urea excretion during the first 6 days and it then returned to normal values at the end of 8 days (fig. 2). At the end of the 8-day experimental period there was no change in the urea cycle enzyme activities of the liver (data not shown) and the serum urea level was normal (fig. 1).

Thus at the end of the 8-day period there was no apparent change in hepatic urea production or output, while the renal PDG activity was elevated 3-fold.

## References


1 Tannen RL: Ammonia metabolism. Am J Physiol 1978;23:F265–F277.

2 Haussinger D, Gerok W, Sies H: Hepatic role in pH regulation: Role of the intercellular glutamine cycle. Trends Biochem Sci 1984;9:300–302.

3 Welbourne TC: Interorgan glutamine flow in metabolic acidosis. Am J Physiol 1987;253: F1076–F1096.

4 Pitts RF: Renal production and excretion of ammonia. Am J Med 1964;36:720.

5 Owens EE, Robinson RR: Amino acid extraction and ammonia metabolism by the human kidney during prolonged administration of ammonium chloride. J Clin Invest 1963;42:263–276.

6 Alleyne GAO: Renal metabolic responses to acid-base changes. II. The early effects of metabolic acidosis on renal metabolism in the rat. J Clin Invest 1970;49:943–951.

7 Haussinger D: Hepatocyte heterogeneity in glutamine and ammonia metabolism and the role of an intercellular glutamine cycle during ureagenesis in perfused rat liver. Eur J Biochem 1983;133:269–275.

8 Phromphetcharat V, Jackson A, Dass PD, Welbourne TC: Ammonia partitioning between glutamine and urea: Interorgan participation in metabolic acidosis. Kidney Int 1981;20: 598–605.

9 Nissim I, Cattano C, Nissim M, Yudkoff M: Relative role of the glutaminase, glutamate dehydrogenase, and AMP-deaminase pathway in hepatic ureagenesis: Studies with $^{15}N$. Arch Biochem Biophys 1992;292:393–401.

10 Ruderman NB, Berger M: The formation of glutamine and alanine in skeletal muscle. J Biol Chem 1974;249:5500–5506.

11 Haussinger D: Nitrogen metabolism in liver: Structural and functional organization and physiological relevance. Biochem J 1990;267:281–290.
12 Meijer AJ, Lamers WH, Chamuleau RAFM: Nitrogen metabolism and ornithine cycle function. Physiol Rev 1990;70:701–748.
13 Jungerman K, Katz N: Functional hepatocellular heterogeneity. Hepatology 1982;2:385–395.
14 Welbourne TC, Phromphetcharat V, Givens G, Joshi S: Regulation of interorganal glutamine flow in metabolic acidosis. Am J Physiol 1986;250:E457–E463.
15 Oliver J, Koelz AM, Costello J, Bourke E: Acid-base induced alterations in glutamine metabolism and ureagenesis in perfused muscle and liver of the rat. Eur J Clin Invest 1977;7: 445–449.
16 Atkinson DE, Camien MN: The role of urea synthesis in the removal of metabolic bicarbonate and the regulation of blood pH. Curr Top Cell Regul 1982;21:261–302.
17 Haussinger D, Gerok W: Hepatic urea synthesis and pH regulation. Role of $CO_2$, $HCO_3$, pH and the activity of carbonic anhydrase. Eur J Biochem 1985;152:381–386.
18 O'Donovan DJ, Lotspeich WD: Factors affecting the deamidation of glutamine by kidney mitochondria of normal and acidotic rats. Enzymologia 1968;35:82–92.
19 O'Donovan DJ, Tobin RB, Horrum MA: The stimulation of the respiration of glutamine by rat renal mitochondria in chronic metabolic acidosis; in Kovacevic Z, Guder WG (eds): Molecular Nephrology: Biochemical Aspects of Kidney Function. Berlin, De Gruyter, 1987, pp 251–256.
20 O'Donovan DJ: Glutamine metabolism in carbon tetrachloride-treated acidotic rats. Int J Biochem 1980;12:85–87.
21 O'Donovan DJ: A comparative study of the renal glutaminase response to acidosis; in Guder WG, Schmidt U (eds): Current Problems in Clinical Biochemistry. 8. Biochemical Nephrology. Bern, Huber, 1977, pp 236–244.

Prof. Daniel J. O'Donovan, Department of Physiology,
University College, Galway (Ireland)

# Subject Index